Solar Technology

The Earthscan Expert Guide to Using Solar Energy for Heating, Cooling and Electricity

David Thorpe

SERIES EDITOR:
FRANK JACKSON

London • New York

First published 2011
by Earthscan
2 Park Square, Milton Park, Abingdon, Oxon OX14 4RN

Simultaneously published in the USA and Canada
by Earthscan
711 Third Avenue, New York, NY 10017

Earthscan is an imprint of the Taylor & Francis Group, an informa business

This book was written using principally metric units. However, for ease of reference of readers more familiar with imperial units, the author has inserted these in the text in brackets after their metric equivalents. Please note that some conversions may have been rounded up or down for the purposes of clarity.

British Library Cataloguing in Publication Data
A catalogue record for this book is available from the British Library

Library of Congress Cataloging in Publication Data
Thorpe, Dave, 1954–
The Earthscan expert guide to using solar energy for heating, cooling, and electricity / David Thorpe. – 1st ed.
p. cm.
Includes bibliographical references and index.
1. Solar energy. 2. Solar buildings. I. Earthscan. II. Title. III. Title: Expert guide to using solar energy for heating, cooling, and electricity.
TJ810.T57 2011
696–dc22

ISBN: 978-1-84971-109-8 (hbk)

Typeset in Sabon
by Domex e-Data, India

Printed and bound by Bell & Bain Ltd., Glasgow

Contents

List of Figures, Tables and Boxes

Figures

Tables

Boxes

Acknowledgements

Thanks to Pat Borer, Katie Brown, Lyne Dee, Frank Jackson, Lightbucket, Pamela Murphy and Adam Paton.

List of Acronyms and Abbreviations

AC	alternating current
BCH	Banco Central Hipotecario
BIPV	building-integrated photovoltaics
BRE	Building Research Establishment
BTES	borehole thermal energy storage
Btu	British thermal unit
CDM	Clean Development Mechanism
CFLR	compact Fresnel lens reflector
CHP	combined heat and power
COP	coefficient of performance
CPC	compound parabolic concentrating
CPV	concentrator solar cells
CSP	concentrating solar power
DC	direct current
DHW	domestic hot water
DNI	direct normal irradiation
DTC	differential thermostatic controller
ELCC	effective load carrying capability
EPDM	ethylene propylene diene monomer
ESTELA	European Solar Thermal Electricity Association
FITs	'feed-in' tariffs
GHI	global horizontal irradiation
HDDs	heating degree days
HDPE	high-density polyethylene
HVAC	heating, ventilation and air conditioning
IEA	International Energy Agency
ISE	Fraunhofer Institute for Solar Energy Systems (Freiburg)
LCOE	levelized costs of energy
LEDs	light emitting diodes
LEED	Leadership in Energy and Environmental Design
LFR	Linear Fresnel Reflector
MD	membrane distillation
MEDISCO	MEDiterranean food and agro Industry applications of Solar COoling technologies
MESoR	Management and Exploitation of Solar Resource Knowledge
MPPT	maximum power point tracking
MVHR	Mechanical Ventilation with Heat Recovery
nm	nanometres
NREL	National Renewable Energy Laboratory
PCM	phase-change material
PEX	cross-linked polyethylene
PHPP	Passive House Planning (Design) Package
PPE/PS	polyphenylene ether/polystyrene
PRINCE	Promoters and Researchers in Non-Conventional Energy
PV	photovoltaic

PVC/PU	polyvinyl chloride/polyurethane
PVG-IS	Photovoltaic Geographical Information System
PV/T	PV power generation and solar water heating
RO	reverse osmosis
SHGC	solar heat gain coefficient (US)
SOLEMI	Solar Energy Mining
SPH	solar pool heating
STC	standard test conditions
SVTC	Silicon Valley Toxics Coalition
SWH	solar water heating
THERM	Two-dimensional building HEat tRansfer Modelling
TISS	Thickness Insensitive Spectrally Selective
UV	ultraviolet
VAT	value added tax
WAPI	Water Pasteurization Indicator
W/m^2	watts per square metre

1 Introduction

> *One must not believe, despite the silence of modern writings, that the idea of using solar heat for mechanical operations is recent. On the contrary, one must recognize that this idea is very ancient and its slow development across the centuries has given birth to various curious devices.* Augustin Bernard Mouchot, at the Universal Exposition, Paris, France, after demonstrating a solar-powered device that made ice (1878).

> *Eventually industry will no longer find in Europe the resources to satisfy its prodigious expansion... Coal will undoubtedly be used up. What will industry do then?* Augustin Bernard Mouchot, after demonstrating an early industrial application of solar thermal energy (1880).

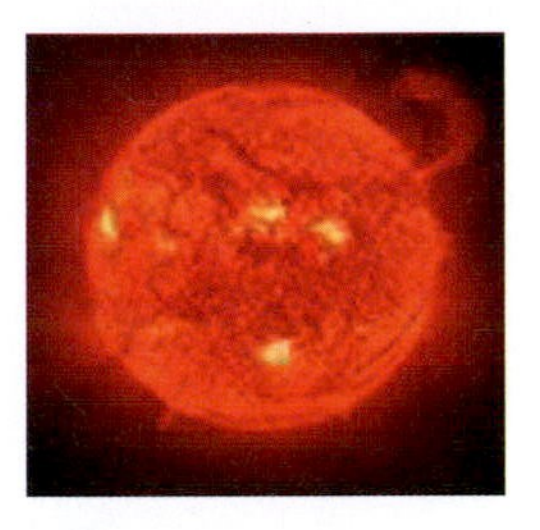

Figure 1.1 The sun.

Source: Wikimedia Commons

Augustin Mouchot was an exceptional pioneer of solar power who invented the first solar-powered engines 150 years ago (Figures 1.2 and 1.3). Since then, our knowledge of how to exploit the power of the sun has increased by leaps and bounds, especially in the last 50 years. Nowadays, with the reality of climate change upon us, it seems that almost every week we are learning about a new way to exploit the incredible amount of solar energy that falls upon our planet every day.

From massive solar power stations in the northern Sahara that may soon help provide Europe with its energy needs, to houses that don't need any heating or cooling (Passivhaus); from a solar-powered iPhone case/charger, to a power station that uses phase-changing materials to let the sun run our homes, even at night; from an electric car with solar panels on the roof to help save fuel, to a solar-powered ski jacket; all of these innovations and many more are coming over our event horizon. Like the rising sun, they herald a potential new dawn for our energy-guzzling world, if only it can wean itself off dirty fossil fuels and high risk nuclear energy that both bring with them massive environmental hazards. Unlike them, the freely available power source of solar energy is everywhere, brings no dangerous pollution, and will last longer than the human race.

The sunshine we receive every day could provide more than enough power for our global needs, even with a bigger population in the future. We can learn to heat and cool our buildings without the need for electric heating or cooling devices, as the Romans did. We can capture the sun's heat in solar water heating panels to provide hot water. We can use its light to make electricity in photoelectric solar panels. We can even copy what nature itself does to make plants grow, by employing its light energy in photosynthesis to provide for our

Figure 1.2 French solar pioneer Augustin Mouchot (1825–1912).

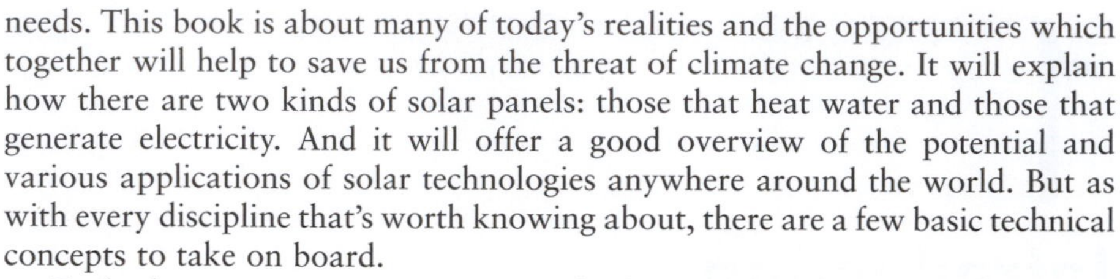

needs. This book is about many of today's realities and the opportunities which together will help to save us from the threat of climate change. It will explain how there are two kinds of solar panels: those that heat water and those that generate electricity. And it will offer a good overview of the potential and various applications of solar technologies anywhere around the world. But as with every discipline that's worth knowing about, there are a few basic technical concepts to take on board.

Each chapter contains a section on the history of each technology. Several thought-provoking ideas emerge from this. First, it's interesting how instrumental the French have been in the development of solar power – from Antoine Lavoisier, Augustin Mouchot, Abel Pifre and Felix Trombe right up to the developers of the new solar-powered aeroplanes and blimps. Even more interesting is the way in which solar technology has proceeded in leaps and bounds over the last three centuries, and at several points been abandoned in favour of coal or oil. This happened with concentrating solar power stations in 1914, and solar thermal water heaters in America in the 1930s, to cite but two examples. It is as if it has constantly been on the edge of a breakthrough as the world's principal source of energy, and if it hadn't been for the abundant supply of cheap oil, it would be much more widespread today. The abundance of fossil fuels in the crust of the Earth has propelled human development in the last 200 years, but has brought us to a point of crisis because of climate change. Consider if there had been less of this fuel bequeathed to the human race: suppose it had, for instance, run out in 1914 – we would now be well into the solar age. Consider again if it had run out in the 1970s – then we wouldn't be arguing now about whether human activities have caused climate change. Whether the solar age finally comes in the next 10 years, or much later in this century, there is no doubt that it will come. And it is to the pioneers named in these pages that future generations will owe a huge debt.

Figure 1.3 On 6 August 1882 this printing press produced copies of *Le Chaleur Solaire* (Solar Heat) by Augustin Mouchot, a newspaper that he created in the Tuileries Gardens, Paris, for the festival of L'Union Francaises de la Jeuenesse. It printed 500 copies an hour, using solar thermal technology.

Source: Corbis

The range of technologies

There is far more solar energy potentially available than we could ever use, but the problem with collecting it on a large scale is that it is not especially concentrated. We would need a big area of the Earth's surface to capture enough to convert it into electricity for all our needs. In fact, someone has calculated it would require around 1 million km^2 (386,102mi^2) of land surface. Matthias Loster at the Department of Physics, University of California, worked out that this amount of land could in theory be deployed at six sites around the globe: the Sahara; the Great Sandy Desert in Australia; Takla Makan in China; Iraq/Kuwait; the Atacama Desert in South America; and the Great Basin of the US (Figure 1.4). Together these installations could produce 18TWe, the world's total primary energy supply in 2006 using current levels of technology. He assumed panels with a conversion efficiency from incident sunlight to electricity of 8 per cent. One project – Desertec – is already under way in the Sahara to

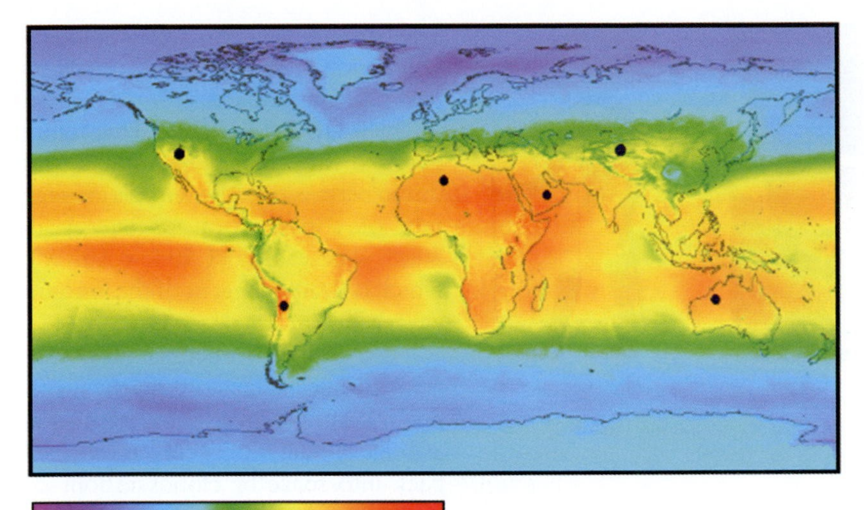

Figure 1.4 It's been calculated that solar power stations in just six of the hottest sites on the world's surface (black dots) could harvest enough energy for all our needs: 18 terawatts. The colours represent the amount of solar energy hitting the Earth's surface in W/m^2 averaged out over the year, indicated by the scale below the map.

Source: www.ez2c.de/ml/solar_land_area © Matthias Loster

build such an installation to supply Europe – except that it doesn't expect to use photovoltaic (PV) technology but concentrated solar power. This focuses direct sunlight onto a type of heat engine that is used to produce electricity.

For solar PV is just one of many solar technologies. This book will systematically examine them all:

- passive solar architecture;
- solar water heating;
- solar thermal electricity generation;
- photovoltaic electricity generation.

For simplicity's sake, this is a not-exhaustive list of the technologies we will cover from the point of view of their end use:

- heating and cooling space: passive solar design, urban planning, passive stack ventilation, phase-change materials, unglazed transpired collectors, solar-powered chillers and coolers;
- lighting: glazing, special glass coatings; sun pipes;
- heating water: solar water heating systems; evacuated tubes; swimming pool heating; active solar cooling; applications for large buildings and districts;
- cooking, food drying, desalination and water treatment;
- electricity: PV modules, system design, process heat, concentrating solar power;
- transport: solar vehicles, hydrogen production.

Solar radiation and the Earth

Those human cultures which placed a solar deity at the top of their pantheons recognized something fundamental: it all starts with the Sun (Figure 1.5). Almost all energy found upon the Earth – upon which all life is dependent – has its origin in that great glowing ball 150 million km (92 million miles) away.

Figure 1.5 Many traditional and ancient cultures around the world acknowledged the importance of the sun as a life-sustainer, including the Greeks, with (a) Apollo and (b) Egypt's sun god Ra.

Source: Wikimedia Commons

The sun – 'Sol' to astronomers – sends us a staggering 174 petawatts (PW) of solar radiation (or insolation) – which is 1.74×10^{17}W. This figure is called the 'solar constant'. To put this in perspective, the world's average rate of consumption of energy in 2008 was 15TW (1.504×10^{13}W), which is 0.008 per cent of the amount we receive.

The sun's energy becomes transformed when it reaches our planet's atmosphere. Some of it – on average about 29 per cent – is reflected back into space by clouds. About 23 per cent is absorbed in the gases of the atmosphere on the way down. Only the remainder – less than half – reaches the surface (Figure 1.6). So although satellites have measured an average of 1366 watts (W) per square metre of solar energy arriving at Earth's front door – a figure called the total solar irradiance – in practice, on average, only about 340W per square metre (W/m^2) reaches the surface.

This average figure is also lower than you would expect (a quarter of the total) because most surfaces on the planet are not perpendicular to (facing) the sun's rays, therefore this energy is spread out over a larger area. Due to the 23.5° tilt of the Earth's axis relative to the plane at which it circles the sun, the only places where it is at its zenith – directly overhead – are in summertime, between the tropics, at midday – called the solar noon (which is not always the same as local time noon).

Box 1.1 Units

kilo-	k	10^3	10,000
mega-	M	10^6	10,000,000
giga-	G	10^9	10,000,000,000
tera-	T	10^{12}	10,000,000,000,000
peta-	P	10^{15}	10,000,000,000,000,000

For example:

milliwatt (mW): 1000th of a watt (W)
kilowatt (kW): 1000W
megawatt (MW): 1,000,000W
gigawatt (GW): 1,000,000,000W
terawatt (TW): 1,000,000,000,000W. In 2006, about 16TW of power was used worldwide.
petawatt (PW): one quadrillion watts: 174 of these reaches the Earth from the sun

Box 1.2 Power and energy

Power is the rate at which energy is produced by a generator or consumed by an appliance. *Unit*: the watt (W). 1000 watts is a kilowatt (kW).
Energy is the amount of power produced by a generator or consumed by an appliance or over a period of time.
Unit: the watt-hour (Wh). 1000 watt-hours is a kilowatt-hour (kWh), commonly a unit of electricity on a bill.
Alternative unit: the joule (J). Watt-hours can be used to describe heat energy as well as electrical energy, but joules are also used for heat. 3600 joules = 1Wh. Put another way, a joule is one watt per second, since there are 3600 seconds in an hour; or 3.6 megajoules (MJ) = 1kWh.

Examples:

- One PV solar panel producing 80W for two hours, or two panels producing 80W for one hour would produce 2 × 80 = 160Wh.
- Three panels producing 90W for five hours will produce 3 × 90 × 5 = 1350Wh or 1.35kWh.

The Tropic of Cancer is the northernmost latitude – 23°26'17" north of the equator – at which the sun reaches 90° above the horizon at its zenith. At this time, the northern hemisphere is tilted towards the sun to its maximum extent on 21 June, when it is winter in the southern hemisphere and summer in the

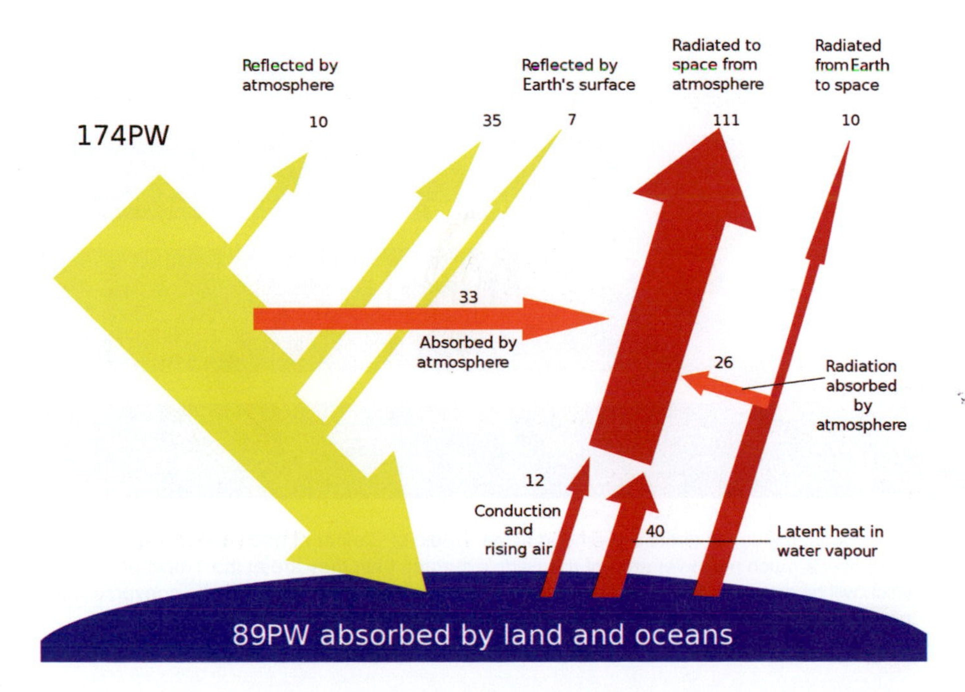

Figure 1.6 The global energy balance. Incoming solar energy is absorbed by different parts of the planetary ecosystem, and most of it is readmitted back into space, otherwise the planet would overheat.

Source: Wikimedia Commons.

northern hemisphere (Figure 1.7). The Tropic of Capricorn is the corresponding latitude south of the equator, and that is directly below the sun at midday on 21 December, when it is winter in the northern hemisphere and summer in the southern hemisphere. The Earth takes one year to circle the sun. On its journey from one extreme to the other throughout the year, the Sun passes over the equator twice – on 21 March and 21 September. These are the two equinoxes (Figure 1.8).

So, the amount of sunlight reaching the ground depends upon latitude and time of day. Figure 1.9 shows how the same amount of sunlight striking a surface at 30° to the horizontal will cover twice the area as that coming from overhead. It might be expected that it would therefore contain half the energy. In practice, it will contain much less, since it is likely to have travelled through twice the depth of atmosphere, which will have absorbed more of its energy on the way.

For this reason, according to the World Energy Council, only half the solar energy that NASA calculates arrives at the Earth's surface is useful. This amount can vary from a maximum of 300W/m^2 in the Red Sea area, to 200W/m^2 in Australia, to 185W/m^2 in the US, and 105W/m^2 in the UK (Figure 1.10). So, the amount of solar power available depends on where you live.

The greenhouse effect and global warming

If no heat escaped from the Earth's atmosphere, it would soon be too hot to live on. Fortunately, energy leaves the planet at more or less the same rate at which

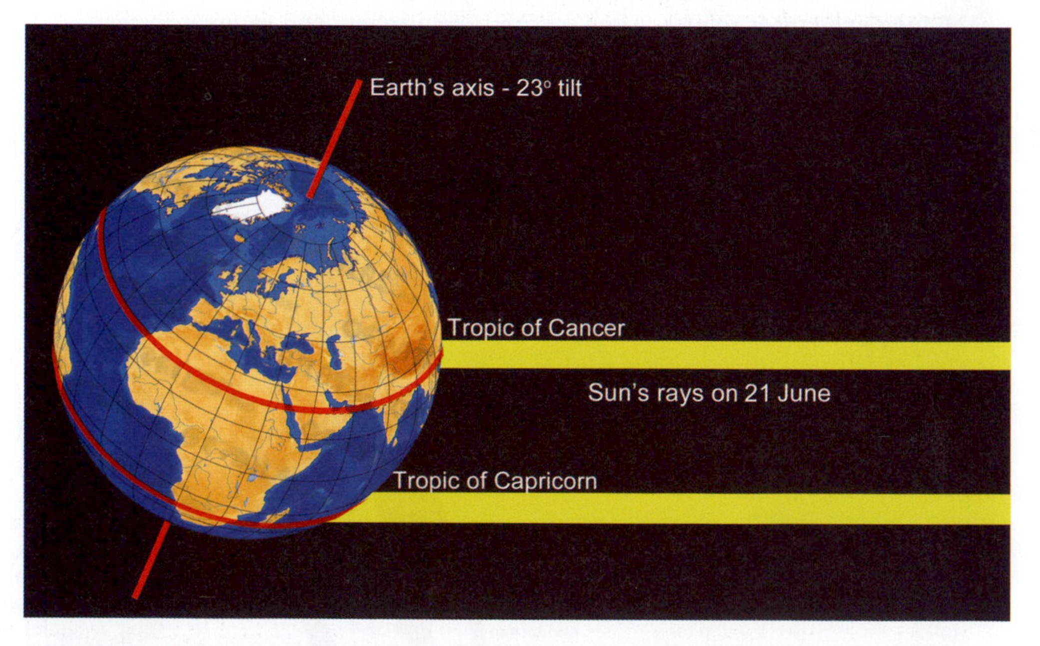

Figure 1.7 On 21 June, the sun is directly overhead at the Tropic of Cancer at noon and its rays are therefore spread over a much narrower area of the Earth's surface than they are at the Tropic of Capricorn (which will be experiencing its winter solstice at this time). Six months later, the converse will be true, as the Earth has orbited to the opposite side of the sun.

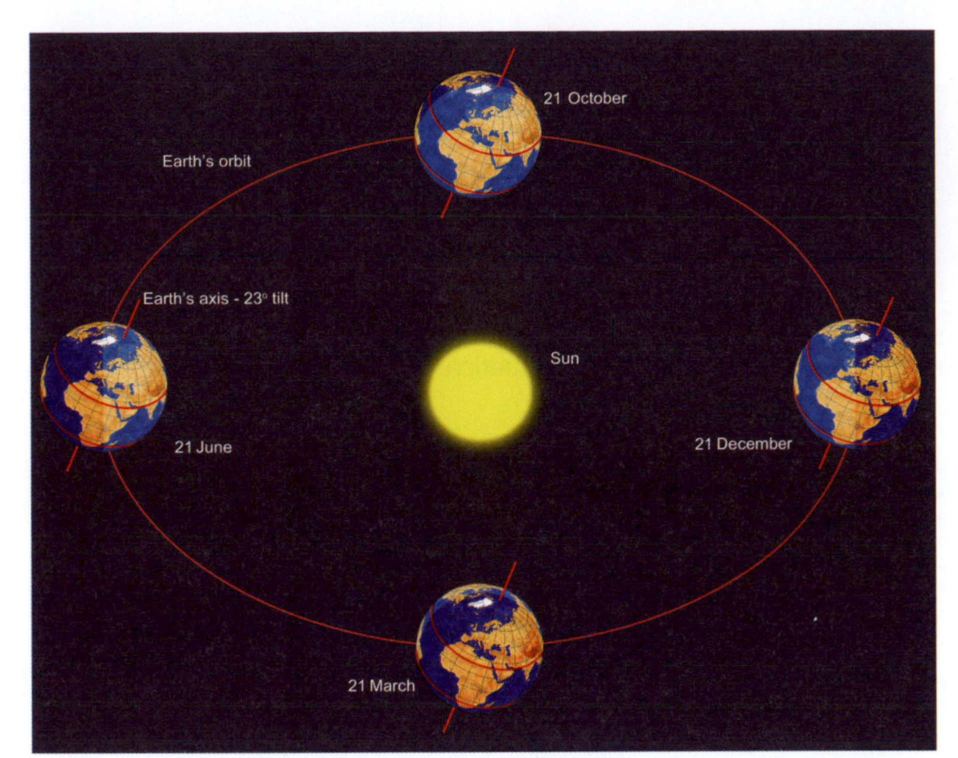

Figure 1.8 As the Earth orbits the sun, its 23° tilt creates the seasons and alters the angle at which its incident rays hit the Earth's surface, and therefore the amount of energy it receives per square metre.

Box 1.3 Solar irradiance

Solar irradiance is the rate at which the radiant energy arrives on a unit area of a surface. It is described in watts per square metre (W/m^2). The incident (arriving) solar radiation is also called insolation and is expressed in terms of irradiance per time unit. If the time unit is an hour ($W/m^2/hr$), then it is the same as the irradiance.

it arrives. This is called 'radiative equilibrium' and means the average global temperature is relatively stable and life can flourish. It happens because certain gases in the upper atmosphere – called greenhouse gases – absorb thermal infrared frequencies and keep heat from escaping into space. The greenhouse gases include carbon dioxide (CO_2), water vapour, methane and ozone. They rise to the top of the atmosphere where they form a blanket-like layer around the Earth, and act rather like the glass in a greenhouse.

Nowadays, the vast majority of scientists accept that concentrations of these greenhouse gases in the upper atmosphere are increasing as a result of humans burning fossil fuels. Consequently, slightly more heat is being kept in the Earth's ecosystem than is leaving.

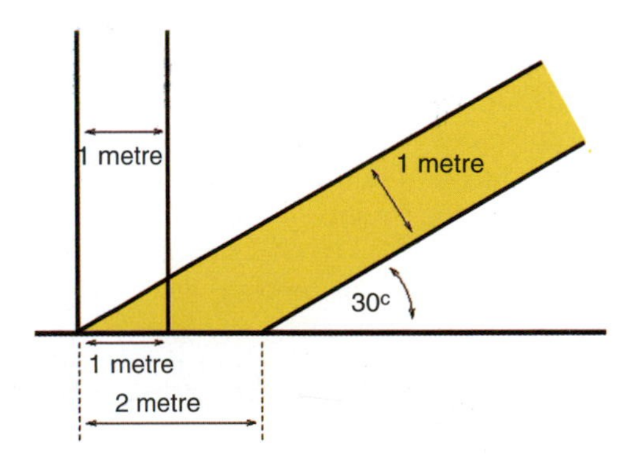

Figure 1.9 How the angle of incidence affects the energy received on a horizontal surface.

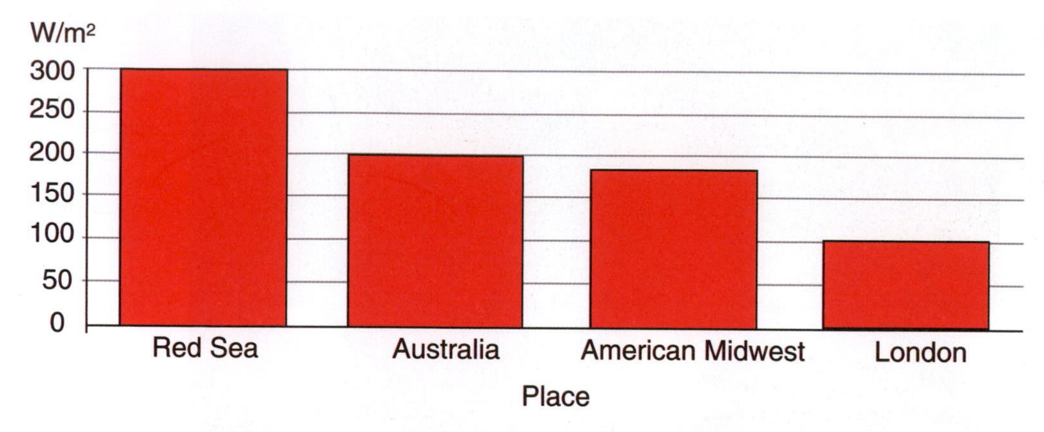

Figure 1.10 Average solar radiation levels at different locations, measured in watts per square metre (W/m^2). For example, the average level of useful solar radiation the American Midwest receives is approximately 180W/m^2. The intensity, of course, varies throughout the year and throughout each day.

Box 1.4 The solar spectrum

The spectrum of solar electromagnetic radiation reaching the Earth's surface is called the solar spectrum (Figure 1.11). It is in the range of 100 to 10^6 nanometres (nm), mostly spread across the visible and near-infrared ranges with a small part in the ultraviolet.

The visible light is used by PV technologies to make electricity. The infrared (heat) frequencies are captured by solar thermal (water heating) and passive building design (space heating and cooling). The breakdown of frequencies is as follows:

- Ultraviolet C (UVC): 100–280nm. Invisible to the human eye.
- Ultraviolet B (UVB): 280–315nm. Responsible for the photo-chemical reaction that produces the ozone layer around the planet.
- Ultraviolet A (UVA): 315–400nm. Used in tanning and therapy for psoriasis.
- Visible: 400–700nm. Used by PV cells to produce electricity.
- Infrared: 700nm–10^6nm (1mm). Used by solar thermal technologies and architecture to capture heat.

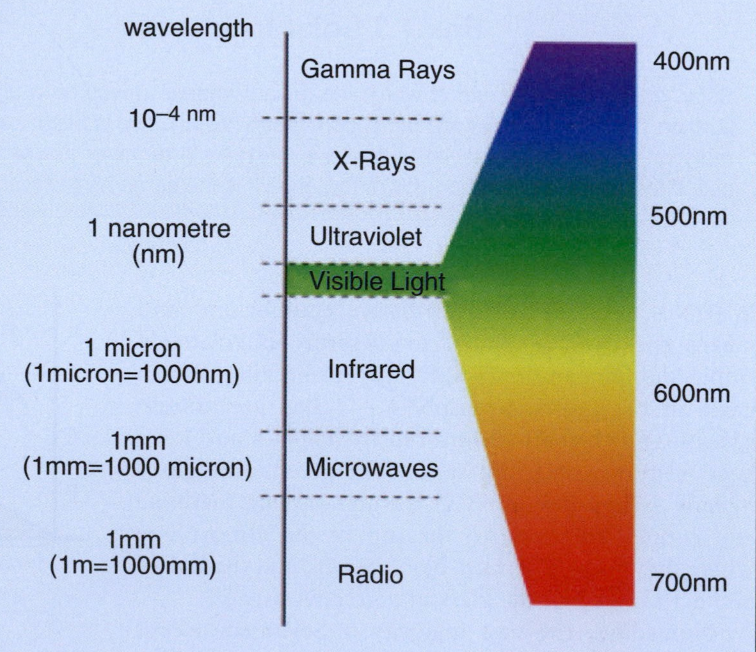

Figure 1.11 The solar spectrum.

Source: Wikimedia Commons

Figure 1.12 The greenhouse effect.

Source: Wikimedia Commons

This is known as global warming, and it is giving rise to climate change, which these scientists agree is resulting in more and more unpredictable and extreme weather events around the globe (Figure 1.12). This is occurring as the historically recent climatic equilibrium is upset by the unprecedented rate of increase of the average global temperature.

How does burning fossil fuels affect this? Photosynthesis – the process by which plants grow – uses sunshine to convert CO_2 in the atmosphere into both carbon that is stored in the growing plants and oxygen that we can breathe.

Fossil fuels – oil, gas and coal – are made from plants that captured carbon dioxide from the atmosphere using this same process millions of years ago. These plants were then buried underground and crushed into coal or turned into oil or methane that was trapped beneath layers of rock. Peat (also burnt as a fuel) contains old vegetable matter too, but it is made much more recently. By taking these fuels out of the ground and burning them, the carbon they contain is released back into the atmosphere, altering the radiative balance (Figure 1.13).

The good news is that we can recapture some of that CO_2 by planting more trees and building with the resultant timber to keep the CO_2 out of the atmosphere (Figure 1.14). And an important reason for the great interest around the world in solar power is because we can use it to replace the burning of fossil fuels. This prevents the emission of more greenhouse gases, and so helps to combat global warming and catastrophic climate change.

Indirect solar energy

A tiny part (about 0.002 per cent) of the solar energy that arrives on our planet is converted into renewable wind energy and tidal energy. This is because the tilt of the Earth's axis – as we have seen – gives rise to the seasons; first one hemisphere and then the other receives more solar energy and longer days. The

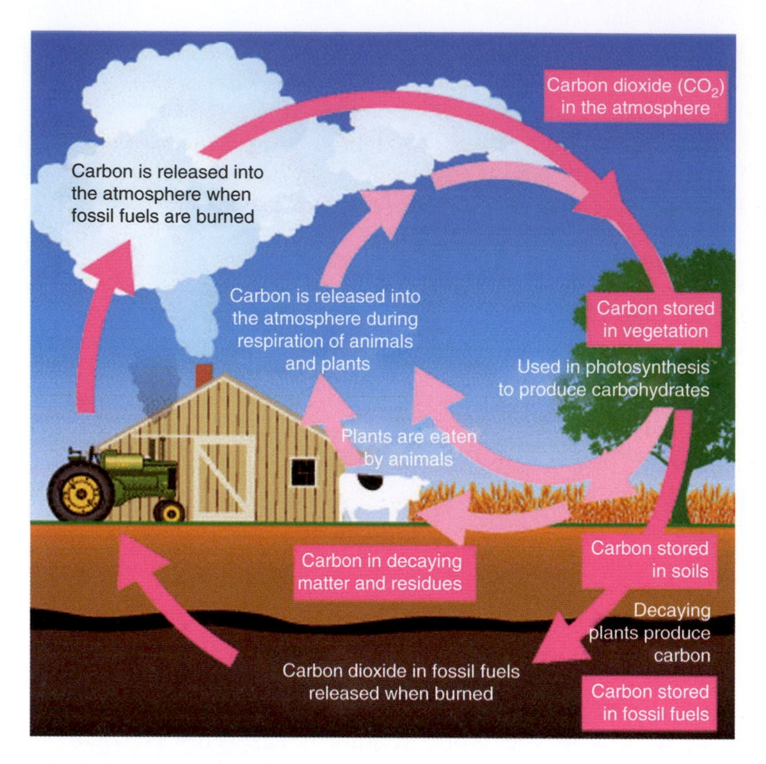

Figure 1.13 The carbon cycle and global warming.

Source: Heather McKee, National Center for Appropriate Technology

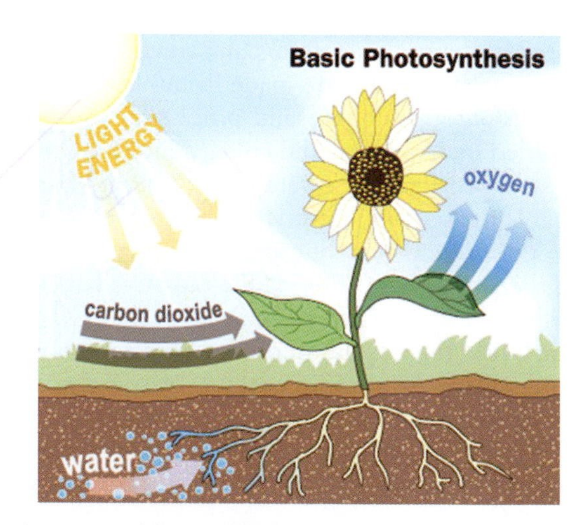

Figure 1.14 The basic process of photosynthesis.

Source: Science Doubts

resulting extreme differences of temperature between different areas of the planet create winds and ocean currents. Evaporation, condensation, convection, air pressure differences and thermal radiation from the Earth's surface all play a part in driving the winds and ocean currents. We can use wind turbines and ocean power devices to capture this immense power. The wind makes and drives the waves; we can capture this energy with wave machines.

Slightly more solar power is converted by photosynthesis into plants and other biomass. We can burn or anaerobically compost this to generate heat and power. As long as these plants are replaced by replanting, then this is also a form of renewable or sustainable energy with a net carbon cost that is almost zero.

Location, location, location

The world is divided into climate zones according to their proximity to the equator, height above sea level and other factors that influence climate (Figure 1.15). These zones receive differing amounts of sunshine, cloud cover, wind and heat. This determines the solar technologies and solar building designs that will be appropriate in these locations. We will consider them here because they are relevant to all of the solar technologies discussed in the following chapters. The principal regions, each discussed in turn, are:

- arid and hot;
- tropical;
- Mediterranean;
- temperate;
- mountainous;
- polar.

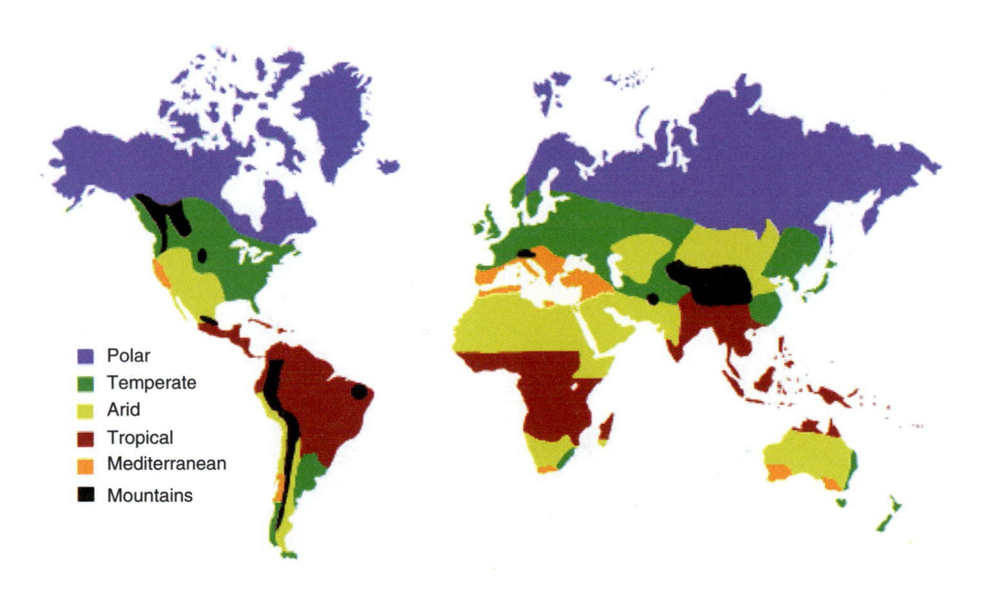

Figure 1.15 The world's principal climate zones.

Source: UK Meteorological Office

Arid and hot, e.g. Egypt

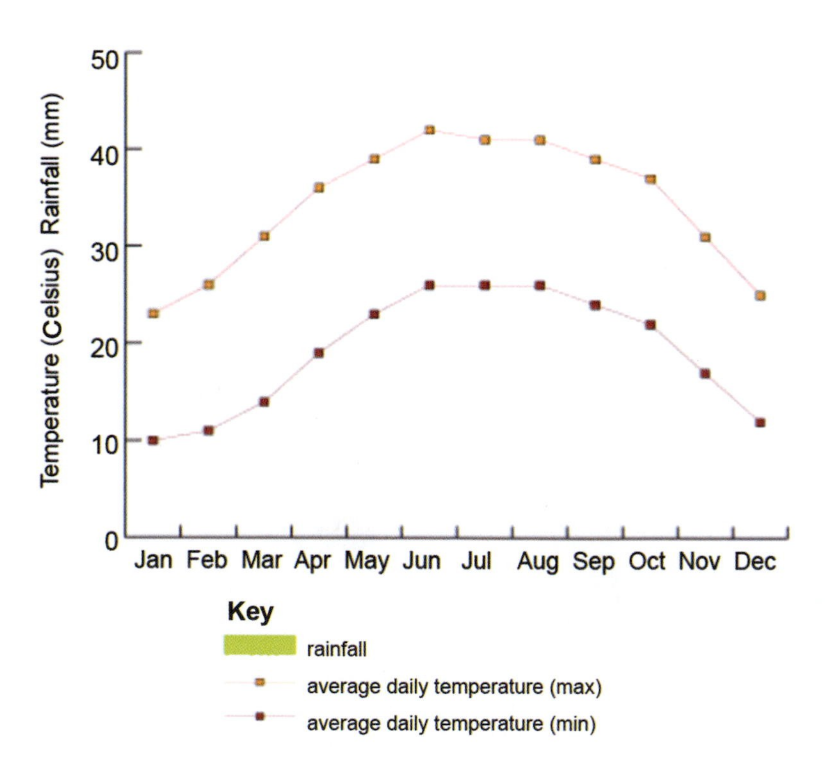

Figure 1.16 Very hot, dry and sunny, with little rainfall all year; the sun is high in the sky.

Source: UK Meteorological Office

Tropical: e.g. Manaus, Brazil

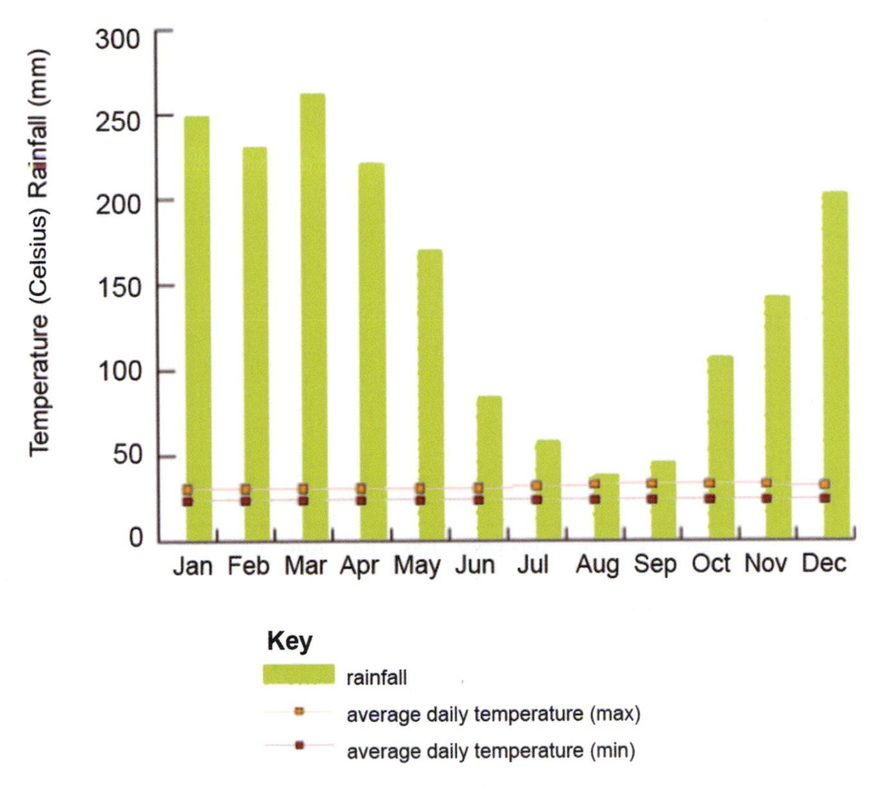

Figure 1.17 Hot, wet and humid all year. The sun is high in the sky.

Source: UK Meteorological Office

Mediterranean: e.g. Miami, Florida

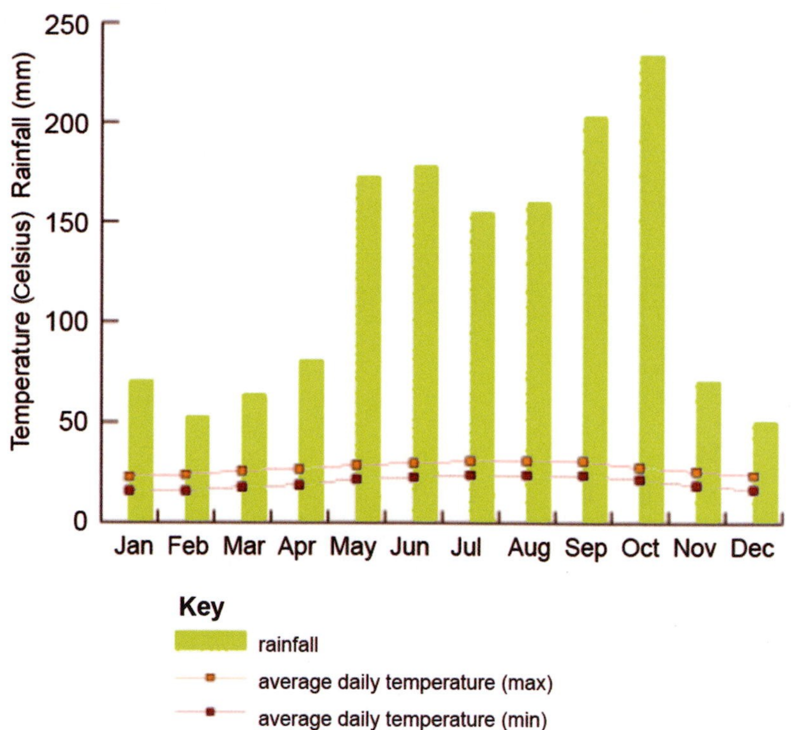

Figure 1.18 Warm and sunny in winter, and very warm in summer, with thunderstorms. The sun is high in the sky in summer, lower in winter.

Source: UK Meteorological Office

Temperate: e.g. Plymouth, England

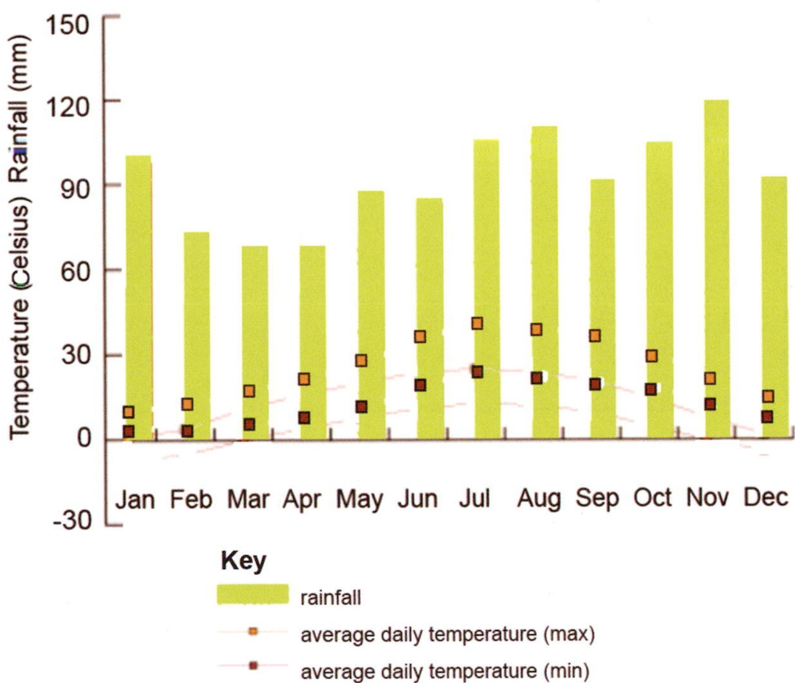

Figure 1.19 Warm summers, cool winters, rainy and changeable. The sun is quite high in the sky in summer, much lower in winter.

Note: Temperate zones are between the tropics and the Arctic or Antarctic, in the middle latitudes between 23°26'22"N and 66°33'39"N, and between 23°26'22"S and 66°33'39"S.
Source: Wikimedia Commons

Mountainous: e.g., Innsbruck, Austria

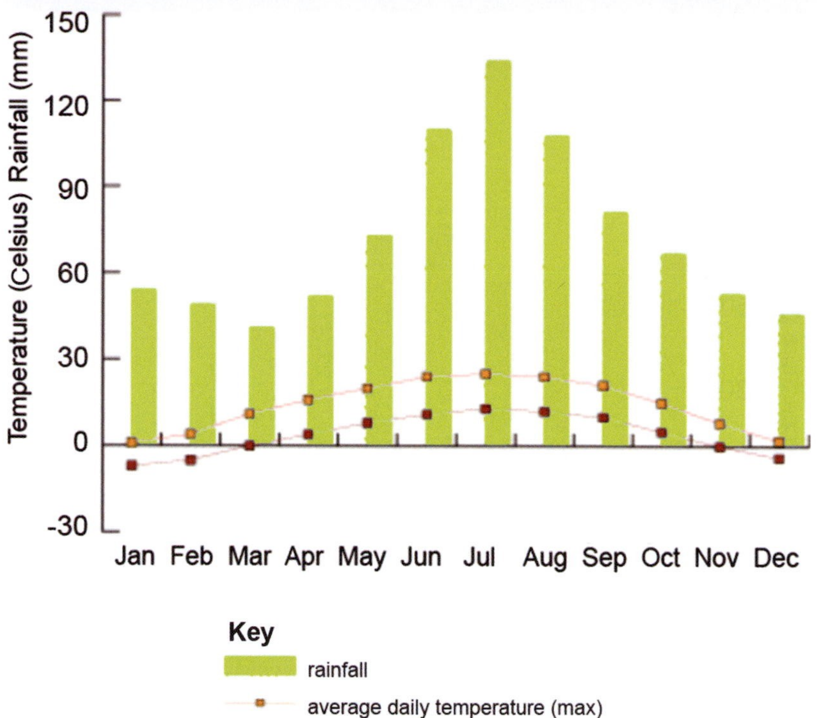

Figure 1.20 Cold and sunny with low temperatures and snow in winter and more cloudy and wet in summer. The height of the sun depends on latitude.

Source: UK Meteorological Office

Polar: e.g., Churchill, northern Canada

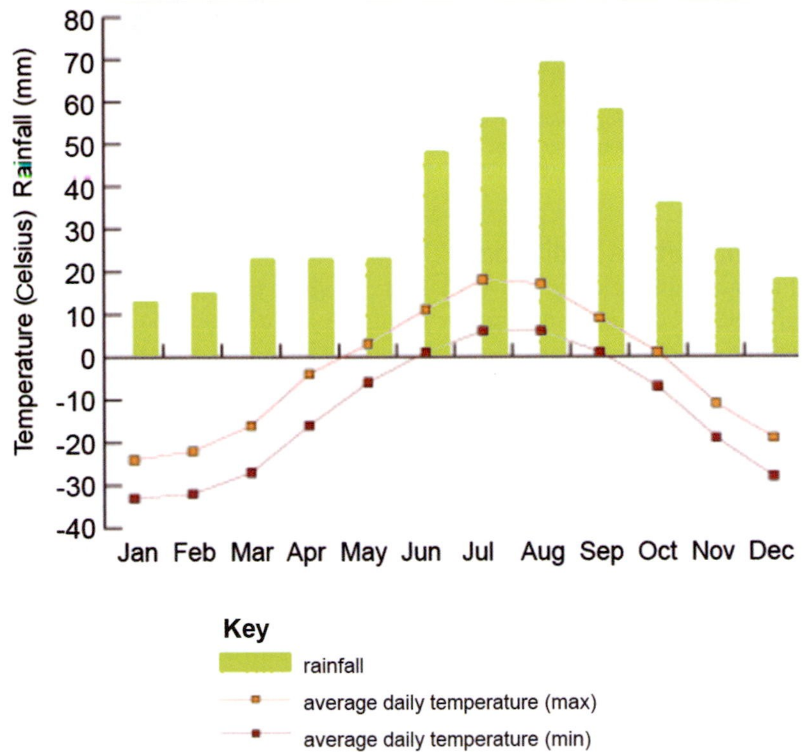

Figure 1.21 Permanent snow and ice in long winters, with very low temperatures that can be warm in a short summer. The sun is low in the sky, especially in winter when it all but disappears.

Source: UK Meteorological Office

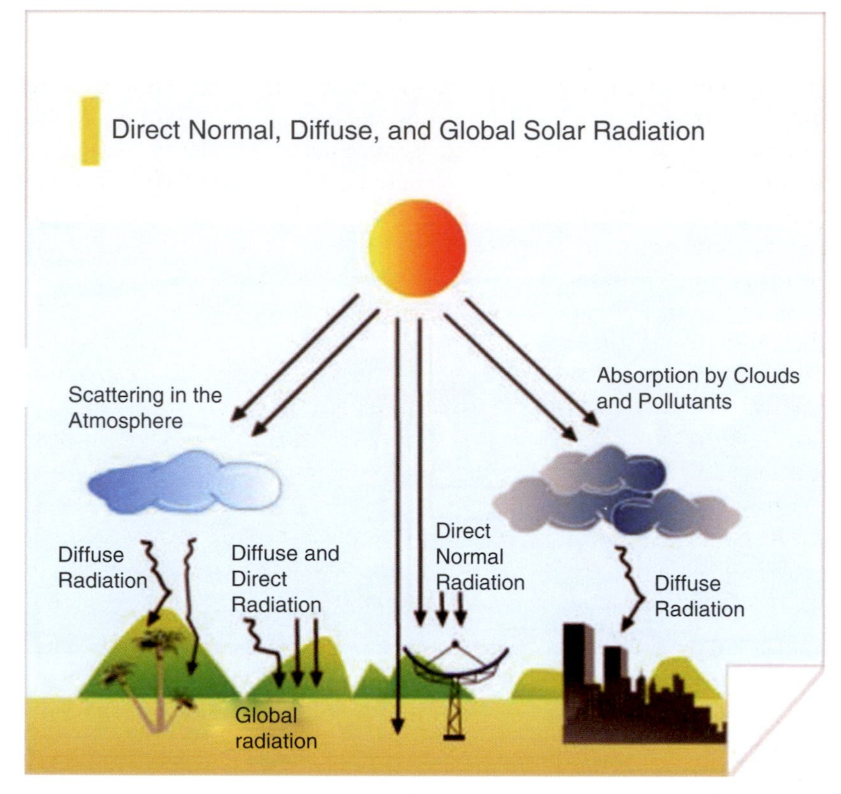

Figure 1.22 Direct radiation arrives on a clear sunny day; diffuse radiation arrives filtered and reflected by clouds, pollution and water vapour.

Direct and diffuse radiation

Another factor influencing the availability of solar energy is how much direct, as opposed to diffuse, sunlight is available at a location. Direct sunlight (sometimes called beam radiation), when the sun is not obscured by clouds, is the most useful form of solar radiation, especially for the purposes of heating (infrared radiation). But daylight is also scattered by particles in the atmosphere. This is called 'diffuse radiation' and it too can be useful for lighting and a reduced amount of PV power. On average, over a year, in northern Europe, half the total solar radiation is direct and half diffuse (Figure 1.22). The term 'insolation' can be used to describe direct, diffuse or total radiation on any orientation of the surface.

Figure 1.23 Measuring solar radiation with an electronic solar insolation meter.

Estimating the amount of energy available

This introductory book is too general to give sufficient information to enable calculations to be made for specific projects at specific locations. There are companion books available in this series that provide this data for off-grid and grid-connected photovoltaics and solar water heating on a small/medium scale, and in Chapter 10 Resources are listed websites where detailed insolation figures can be found (see Figure 1.23). A wealth of data is available to enable calculations to be done at specific locations all around the world. These official figures are based on years of careful observation.

The following instruments are used to collect data over years; they are also used to measure performance of a system against solar radiation. Developers will conduct site specific surveys using pyranometers (which measure the total irradiation), pyrheliometers (which measure only the direct solar radiation), or a Campbell-Stokes sunshine recorder (which only measures the number of infrared insolation hours). These sensors will be tilted to match the tilt angle of the modules or collectors to be installed. The diffuse component can be broken down into an isotropic part, received from all of the sky's dome, a circumsolar part, brighter around the sun's disc, and horizon brightening. The information is fed into functions that take into account variables, such as the azimuth (the direction of the sun, expressed as the angular distance from the north or south point of the horizon to the point at which a vertical circle passing through the object intersects the horizon), latitude, seasonal variation, climate, orientation and so on.

The Resources section also contains references to software, available both freely and at a price, for making calculations based on specifications for systems, buildings, materials and insolation data.

Figure 1.24 A parabolic solar thermal dish in Arizona, US, focusing direct infrared heat at around 1000°C (1832°F) onto a Stirling engine, to generate electricity.

Source: Courtesy of Stirling Energy Systems, Arizona, US

Conclusion

In Chapter 4, in the section on concentrating solar power, the world's first solar power station is described. I tell the story of its success and readers may be surprised by the date of its operation – 1913. I reveal how it was unfortunately abandoned as the First World War plunged the globe into chaos. This war was partially about oil. The so-called 'Great Powers' fought over access to the newly discovered oilfields near Kuwait. Its causes and outcome have much in common with many conflicts that have happened since then, not least the much more recent first and second Iraq Wars. Nowadays, the case for renewable energy is being made in the name of energy security. But only a complete abandonment of oil and all other fossil fuels as a source of energy, and a return to decentralized energy, of a nature that is appropriate for every given location, will give us true global energy security and an end to wars based on competition for limited resources. Why argue over access to something as abundant as sunshine? Especially when it also tackles the other tragic consequence of our rush to depend so heavily on fossil fuels: climate change.

Solar power is a truly exciting field that is rapidly developing, and in the final pages we look briefly at emerging technologies on many different fronts (e.g. Figure 1.24). Globally, the growth of innovative solar technology is happening at a faster rate than any other energy technology. It needs to, because global energy demand is also rising quickly. The global growth of PV power installations alone is just about keeping pace with the global growth of energy demand. With the other solar technologies, it is possible that over the next 50 years solar power can meet a much higher percentage of demand and, together with the other renewable energy sources, gradually replace fossil fuels. When technologies dependent upon these fuels finally come to seem as archaic as a steam engine does to us today, we will know that we are finally living in the solar age.

References

BP, June 2009, 'BP Statistical Review of World Energy 2009', bp.com/statisticalreview

Lindsey, Rebecca, 2009, 'Climate and Earth's Energy Budget', NASA, available at http://earthobservatory.nasa.gov/Features/EnergyBalance/

Loster, Matthias, 2006, 'Total Primary Energy Supply: Required Land Area', www.ez2c.de/ml/solar_land_area/

WEC, 2007, *Survey of Energy Resources*, www.worldenergy.org/publications/survey_of_energy_resources_2007/solar/720.asp

2

Passive Solar Building

Passive solar building design is the oldest use of solar power. It maximizes the direct use of solar power (without employing a panel or module) for lighting, heating and cooling in order to minimize the use of imported forms of energy. It means that elements of the building construction themselves collect solar energy and manage it efficiently. By contrast, active solar systems use added-on panels, collectors and stores, with pumps and fans. In practice, this distinction can become blurred and the two are often combined. However, keeping in mind the separate principles helps a designer minimize artificial energy use.

A certain amount of the sun's heat and light needs to be allowed in and retained within the building. The quantity will vary throughout the year according to the seasons. Passive solar architecture – sometimes called 'low-energy building design' – should allow for the successful management of the internal lighting and air quality, the humidity and temperature, depending on the outside conditions. Many activities within a building create heat: our bodies give off heat, as does cooking, using hot water and electrical equipment, such as printers, computers and refrigerators. If managed properly, these sources of heat, together with that of the sun itself, could be all that is needed in order to maintain a thermally comfortable internal environment. A certain minimum number of air changes per hour should be permitted to keep the atmosphere fresh. Draughts or ventilation should be controllable in order to prevent overheating or cold temperatures, depending on the conditions outside.

Box 2.1 Elements of successful passive solar design

- attention to building form and orientation;
- use of high-performance windows/glazing;
- use of thermal mass to avoid overheating by day and to release stored heat by night;
- high levels of insulation and airtightness;
- user-controllable heating, cooling, ventilation and daylighting;
- installing efficient systems to meet remaining loads;
- ensuring that individual energy-using devices are as efficient as possible, and properly sized;
- ensuring the systems and devices are properly commissioned and maintained;
- educating users in how to successfully occupy the buildings.

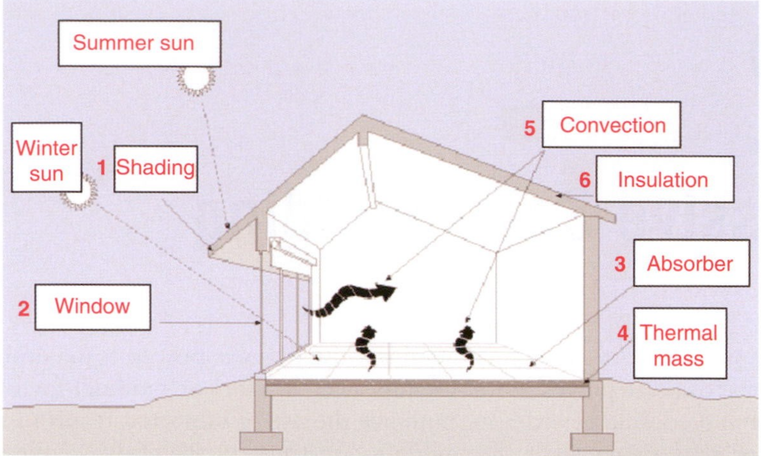

Summer sun
Winter sun
1 Shading
2 Window
3 Absorber
4 Thermal mass
5 Convection
6 Insulation

Figure 2.1 The six elements of passive solar building design.

Source: UK Crown Copyright

Figure 2.2 Sunlight falling directly on exposed thermal mass. Notice the lack of floor and wall covering, which is important to maximize absorption of the solar energy. This is in a Canadian passive solar house, ÉcoTerra™.

Source: www.solarbuildings.ca/en/main_53

Five basic principles

Exactly how the ideal utilization of solar power is accomplished depends on the location of the building, but regardless of this, many of the principles are the same. There are five basic elements:

1. The light must pass through a window that can be double- or triple-glazed glass, and given a low-emission (low-E) coating to allow light to pass through and reflect infra-red wavelengths back in. Windows will be sized and positioned according to the direction they face relative to the sun's path. There will be an optimum balance between the light required to illuminate the room without glare and minimize artificial lighting, and that required to provide sufficient heating and prevent overheating. This balance will be controllable, as it will vary according to the season, time of day and weather. Light will be directed deep into the building wherever possible.
2. The sunlight must fall onto a surface that will absorb the heat and transfer it into the thermal mass behind it. Rugs or curtains should not therefore be allowed to cover this absorption surface. Thermal mass is the product of the heat capacity of a material and its total mass and conductivity. The more dense a material is, like stone, the more heat it will hold. This means that as it warms up, it will help stop the room overheating and take longer to cool down, keeping the interior warmer during the night. The more dense it is, the longer it will take to warm up from cold and cool down afterwards.
3. Shading outside the window will prevent too much sunlight coming in when the sun is high in the summer, but permit it to enter when it's needed in the winter. Shading can be natural (deciduous vegetation), or involve shutters and blinds that can be manually or electrically operated.
4. Heat rises, and this can be used to help with ventilation and to transport heat by convection from one part of the building to another. The process is sometimes called the 'stack effect'.
5. Insulation and draught proofing will provide a tight thermal envelope (see page 46). Insulation is a heat barrier: it keeps the heat out of a building if we want to keep it cool as well as maintaining heat inside if we want it to be warm.

The perception of heat is also relative. The interior of an underground cave in the winter in a country at a high latitude feels warmer than it does outside. The same cave visited in the summer when it is hot outside will feel cool. Actually, the air temperature varies only slightly underground. We have a clue here then about how to keep a building cool in a hot country: have very few window openings that face the sun, and very thick walls (or at least well-insulated ones).

Conversely, even in a cold climate, when the sun is shining, it can feel hot behind south-facing windows, protected from the cooling effect of wind or cool air. If we can capture this heat by heating up a floor or wall with a high thermal mass, which faces the window (preferably a darker colour to absorb more heat), or to melt a phase-change material, such as wax, then it will give off that heat later, back into the room, when the sun has disappeared. We could also have a vent high up in the wall opposite the window that uses the stack effect to let the hot air rise into cooler and higher parts of the building.

Figure 2.3 A wall or floor with high thermal mass behind a sun-facing window or conservatory can absorb heat in the daytime that is released at night time to keep the interior warm.

The meaning of 'passive' in 'passive solar' may now be becoming clearer. Buildings achieving the high energy savings that are possible can cost no more to build than conventional buildings, due to the downsizing of heating and cooling equipment, and they are superior in other respects as they offer a healthier and more pleasant living environment. Lifetime running costs will always be considerably lower.

A holistic system approach

Passive solar building design must take into account all of the elements of the building and its location as a complete, dynamic system. The design principle should be 'only use energy when you have exhausted design'. For a design team, it could be counter-productive to have somebody other than the overall designer specify an air conditioning system, for example, as it could risk overspecification. The same would apply to a building's windows, which could have inappropriate glazing transmittance characteristics, motorized shades or vents, all of which could work against the efficient operation of the building as an entire system. Although in a simple building it may be unnecessary, by using computer modelling software we can fully optimize the overall design of a building and its mechanical systems to push energy savings to 60–75 per cent. It is possible to enjoy energy savings of 80 per cent and beyond only by enlisting the enthusiastic cooperation of the occupants, which implies some form of training and user manual.

By modelling the entire building's performance and adjusting variables like orientation, window size, overhangs and so on, we can use an iterative process to figure out the best performing compromise between the options. The design should consider lifecycle impacts from conception to demolition. A chief design goal is to satisfy the occupiers' needs while minimizing the total carbon content of the energy used. The best metric to measure this is the cost per tonne of carbon dioxide saved. The carbon impact throughout the lifetime of a building comprises the carbon in the energy used:

1. in the materials, sometimes called 'embodied energy';
2. to refurbish and maintain the building;
3. to run and live in the building;
4. to dispose of the building at the end of its useful life.

Minimizing environmental impact

The embodied energy of fossil-carbon materials may be as high as 50 per cent of the total energy used in a building's lifetime. At the time of installation, the global warming cost of their manufacture is already present in the atmosphere in the emissions thus caused. However, at that point, the fossil energy saved by their insulation value in a building is only potential. In other words, their climate-protecting benefit is yet to be realized over the 100-odd years of the building's life. This is why we must try to use materials such as timber that 'lock up' carbon in the fabric of the building, at least for the time being.

Hemp-lime is a composite concrete of lime mixed with hemp fibre while limecrete is a lime concrete or render with a mineral aggregate. Both have low embodied energy whereas concrete has a very high one. In addition, concrete has a huge impact on global warming due to its sourcing and processing. On the other hand, hemp-lime locks up in the building the carbon dioxide that the hemp absorbed from the atmosphere while it was growing. So, concrete has a net negative effect and hemp-lime composite, along with other renewable materials, has a net positive effect (if sourced sustainably). Limecrete and hemp-lime composite have a much lower structural strength than concrete, but

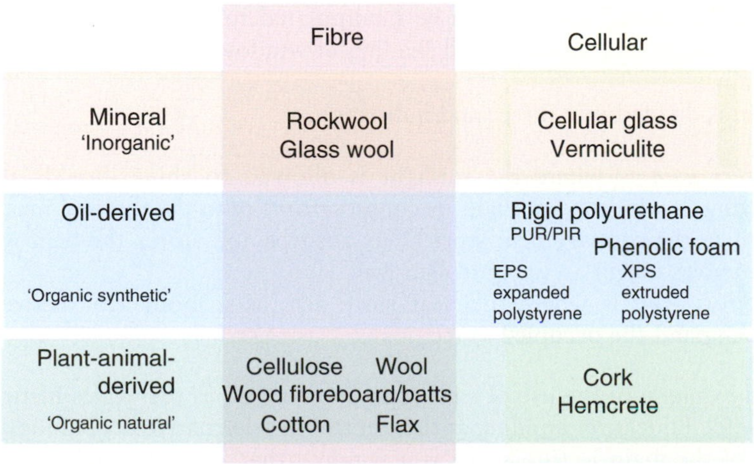

	Fibre	Cellular
Mineral 'Inorganic'	Rockwool Glass wool	Cellular glass Vermiculite
Oil-derived 'Organic synthetic'		Rigid polyurethane PUR/PIR Phenolic foam EPS expanded polystyrene XPS extruded polystyrene
Plant-animal-derived 'Organic natural'	Cellulose Wool Wood fibreboard/batts Cotton Flax	Cork Hemcrete

Figure 2.4 Classification of insulation materials.

do last longer, breathe, are more flexible (and less prone to cracking), and also possess reasonable thermal mass. These renewable, natural materials that lock up carbon should therefore be used where possible.

In terms of insulation cellulose (such as recycled newsprint), straw, wood fibre and other natural materials carry the least embodied energy and have small marginal payback times. Fibreglass or rockwool have more embodied energy, but significantly less than foam insulations like EPS, XPS, spray foam and so on, and therefore would be preferred whenever possible. It so happens that most natural materials, and metals, are also easier to process and recycle at the end of their useful life. Many will degrade or decompose naturally, or can be reconstituted and re-used. Materials should also be sourced as close to the site as feasible; by and large, the embodied energy of a material rises in proportion to the distance it has travelled.

Indoor air quality is another factor that affects materials choice. In general, natural materials off-gas less. PVC use should also be avoided. In February 2007, the Technical and Scientific Advisory Committee of the US Green Building Council (USGBC) published a report for the Leadership in Energy and Environmental Design (LEED) Green Building Rating system which concluded that the 'risk of dioxin emissions puts PVC consistently among the worst materials for human health impacts' (see www.usgbc.org/ShowFile.aspx?DocumentID=2372).

Solar gain

The heating effect of the sun in a building is called solar gain. It varies with the strength of the sun, its angle and the effectiveness of the glazing to transmit or reflect its energy, and the thermal mass of the building to absorb it. Usually, we want to maximize solar gain within the building in the winter (to reduce space heating demand) and to control it in summer (to minimize cooling requirements and solar glare).

Good building design uses layout, orientation, windows and exterior doors containing glass to minimize the need for artificial lighting and to capture within the building the proportion of the sun's heat required to keep a comfortable temperature. The choice of glazing, shading, thermal mass and other factors help to achieve this. The composition and coating on the inner faces of the glazing cavities can be manipulated to optimize the greenhouse effect, while the size, position and shading of windows can be used to optimize solar gain.

Gain is divided into direct and indirect:

- *Direct gain* is where the sunlight is allowed to shine directly through glazing (a window, rooflight or conservatory) onto the thermal mass being used as the primary heat store. This absorbs and stores the heat which it then gives out in an uncontrolled way.
- *Indirect gain* is where the heat store acts as a moderator between the glazing and the occupied space.

A good example of the use of solar gain can be seen in the Wales Institute for Sustainable Education building at the Centre for Alternative Technology. In the ceiling of the main lecture space is a window that looks like the oculus of a

giant telescope, over which a cap can swivel to let in daylight when required. The large building is relatively airtight, with three air changes per hour. Part of the roof is covered with 70m^2 (754ft^2) of evacuated solar water heating tubes. These contribute to the hot water requirements, which are also met by biomass boilers. All of the electricity is also renewably sourced; on an adjacent building, there are 6.5kWp of PV panels.

The stack effect

Natural, 'passive stack' ventilation occurs via convection. This relies on the stack effect – that warmer air is more buoyant than colder air. Warm air rises up through the building to exit at the top through open windows, vents, chimneys or leaks. This suction causes reduced pressure lower down in the building, drawing colder air in through any openings – intentional or otherwise. Good passive solar design takes advantage of this principle to circulate heat and create ventilation. In a high-rise building with a well-sealed envelope, the pressure differences can become acute via stairwells and shafts. Steps should therefore be taken to mitigate it with mechanical ventilation, partitions, floors, and fire doors to prevent the spread of fire.

Figure 2.5 The Wales Institute for Sustainable Education building at the Centre for Alternative Technology in Wales, makes use of passive solar design for some of its space heating. For example the main lecture theatre, a circular space with a thick wall made of rammed earth, is surrounded by a large window on its south and west side – forming a corridor – which allows the sun to heat up this large volume of thermal mass. This heat is then released at night. The rammed earth aspect of the design is similar to that of the Chapel of Reconciliation in Berlin, Germany.

The importance of climate

As we saw in Chapter 1, the world is divided into climate zones. Each requires its own type of architecture in order to be energy efficient and provide comfort for its occupants. For reasons of space, we will confine ourselves to examining

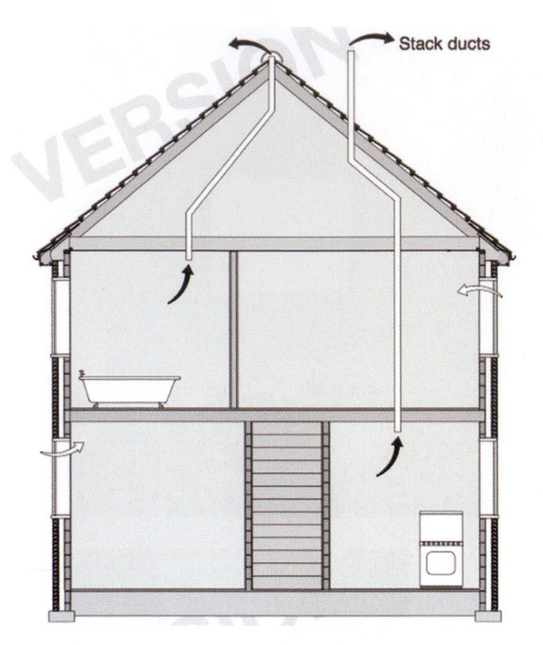

Figure 2.6 Illustration of the stack effect drawing warm, humid air from the 'wet rooms' of a dwelling through the loft, to be vented out via the roof.

Source: © EST

Figure 2.7 Solar chimneys, which use the stack effect to induce natural ventilation, on a building at the Building Research Establishment (BRE), Garston, UK.

Source: BRE

the passive solar and cooling techniques in just two general climates: hot and cool. In transitional climates, elements of each may be combined in the same building as appropriate.

Hot climates

We can learn a lot from the planning and architectural practices of peoples who have lived for centuries in hot climates. Whether it is the ancient Romans, the Egyptians, the Iranians with their windcatching towers, or the Anasazi, who built houses such as the Cliff Palace at Mesa Verde, they all used similar principles. Although their solutions can be very specific and include detailed observations about local wind patterns, caves and rock formations, underground aquifers – all of which can be harnessed for cooling purposes – the principles are the same. They involve attention to:

- spatial planning;
- building orientation;
- thermal mass;
- window position and size;
- external wall coverings;
- shading and overhangs;
- passive ventilation;
- phase change materials.

Nowadays, refinements have been made that can create indoor environments that are pleasant and tolerable without requiring electrical energy to keep them cool. Even if this is not possible, all these techniques should be employed first before using artificial ventilation. It is worth noting that occupants generally modify their behaviour according to the heat, whether it's by wearing appropriate clothing, sleeping in the hottest hours, or moving to the cooler rooms in summer.

Spatial planning and layout

Suffering from oppressive summertime heat and humidity, sprawling Abu Dhabi is an example of how not to design a city in a hot climate. Instead, towns should ideally be compact, like ancient Arabian cities. Streets would be narrow to provide shade and wind towers may be built, drawing draughts through buildings without using energy. These and the

Figure 2.8 Traditional climate-appropriate architecture: Cliff Palace at Mesa Verde, Colorado, US. The overhang provides shade, the huge thermal mass of the rock and the thick walls moderate the internal temperature, while the small windows permit little heat inside but sufficient light.

following characteristics in spatial planning will all help to keep temperatures lower in a block or town:

- reducing the space between buildings to keep the sun from ground level;
- using small courtyards to prevent the sun from reaching the ground floor but allowing heat from the thermal mass of buildings to escape at night;
- using larger courtyards, which will heat up more in the daytime but cool down more at night;
- planting plenty of vegetation in the streets and courtyards to provide shade and evaporative cooling with deciduous trees to interrupt the summer sun's path;
- reducing expanses of Tarmac and concrete, as this can reduce the air temperature above by 20°C (70°F).
- using surface water, ponds, waterways, etc. to provide evaporative cooling.

Building exteriors

The following features for exterior walls and roofs will help:

- exposed surfaces to be whitewashed or a pale colour to reflect heat back at the sky;
- overhangs calculated to keep windows in the shade during summer months but allow in sunlight in winter;
- the building should be wider on the east–west axis than on the north–south axis – this is to minimize solar gain that penetrates beneath the overhangs in the morning and afternoon;

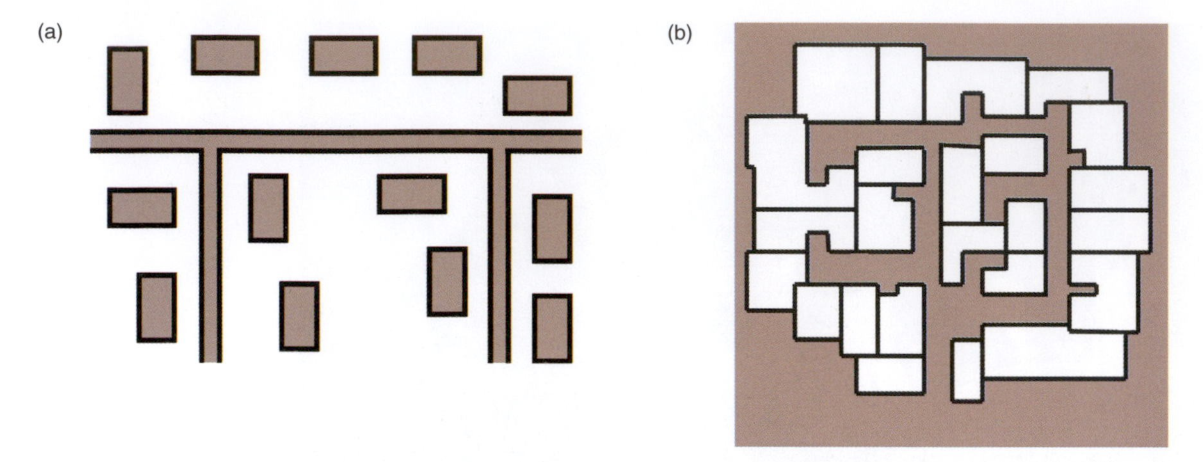

Figure 2.9 (a) Inappropriate and (b) appropriate spatial layouts for towns with hot climates. Grid layouts, and wide spacing of buildings, do not provide shade or wind shelter. Organic, non-grid layouts provide shade and can be designed to block winds, preventing issues with wind funnelling.

Figure 2.10 How not to design a city for hot climates. Abu Dhabi, UAE, shrouded in pollution and a heat haze; glass and steel structures rise up into the sky, perfect collectors of solar radiation, and as a result the buildings require extreme high-energy air-conditioning.

Source: © Frank Jackson

- use of green roofs, arbours and trellises on patios and courtyards – plants provide shade, humidity and cooling;
- few and narrow windows on the sides of buildings which face the sun (east and west);
- walls facing towards the north (in the northern hemisphere) and south (in the southern hemisphere) may have larger windows to let in light;

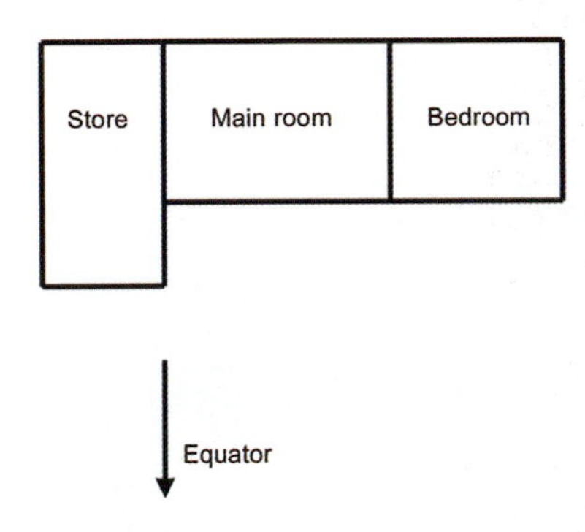

Figure 2.11 Bedrooms need light in the morning and the whole building needs to be protected from low angle evening heat so this is an optimum dwelling layout.

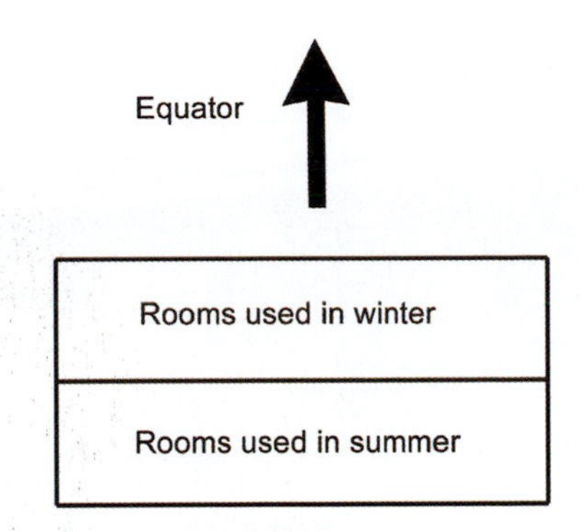

Figure 2.12 Often in hot climates, different parts of a building are used in the winter and summer, as equator-facing rooms become too hot in summer.

- avoid skylights and roof windows to keep out the midday sun;
- high thermal mass for all walls, in particular those facing the equator and east–west;
- shutters and curtains kept closed in the day and open at night;
- external insulation – without insulation, but with thick walls, it will still be cooler in the building and there will be a lag between the temperature outside and inside.

In the US region of Sacramento, California, a west-facing window of just $5m^2$ ($55ft^2$) will add as much as 16kWh (55,000Btu) to a building on a summer afternoon. Running an air conditioner to compensate may require almost 2kW, and high bills, which could be avoided by either proper orientation of the building, or the use of an exterior shade for a fraction of the cost. The early settlers to the area knew the answer to this problem: they usually built their haciendas as south facing, creating long rectangular buildings with porches along this side. In the summer, when the sun was higher in the sky, these helped to shade the buildings and keep them cooler, but in the winter when the sun was lower, it let the light in.

The thickness of the walls is another determining factor in how long it takes for heat to be transmitted from the outside to the inside. Insulation and radiant barriers can also help to keep heat out. This should be on the exterior face of walls.

Shading

The purpose of shading is to prevent solar heat from getting inside the building. However, light should be permitted entry during winter months when it might be needed. Diffuse light should be admitted in order to reduce dependency on artificial lighting. At the same time, natural ventilation is to be encouraged. Combining these aims using shading involves a balancing act.

Figure 2.13 A house in Abuja, Nigeria, with a tin roof and air conditioning installed: an example of how not to tackle the problem of solar heat in sunbelt areas.

Source: © Frank Jackson

Figure 2.14 Traditional shutters and thatch with overhang for shading, in Nigeria.

Source: © Frank Jackson

Many types of shading are possible, and the choice depends on the prevailing wind as well as light. Shade can be provided by nearby buildings, plants, trees and architectural elements. These objects can also be used with wind. They can provide shelter, filter dust (trees) and block wind or direct it to where it's needed to provide ventilation/cooling. Architectural shading elements can be fixed or adjustable. Fixed ones include overhangs, pergolas, vertical fins, balconies, false roofs and, inside, light shelves and louvres. Vines can be grown over

Figure 2.15 Medieval shading from solar heat – an arcade in St John Lateran, Rome, Italy.

Source: © Clare Maynard

pergolas providing a cooling effect. Adjustable elements can include awnings, shutters (inside and out), blinds, rollers and curtains. Shutters are common throughout the Mediterranean. External shades are about 35 per cent more effective than internal ones because they stop the heat getting inside. The lighter the surface of these barriers, the more light they reflect.

External shading should not interfere with night time cooling. For this reason, for example, it is better to have a cover of deciduous plants or creepers than it is to have a concrete wall or galvanized iron sheets on a roof. Heat radiation must be allowed to escape into the cool night sky. Some traditional buildings cover the entire roof surface with small closely packed inverted urban ceramic pots. A removable canvas roof mounted close to the roof during the day and rolled away at night, which can operate automatically, is another cheap solution in mass, traditional housing.

Figure 2.16 Trees can be planted to provide shade in summer while letting in low-angled winter light. Deciduous trees will lose leaves in winter and allow more light and heat to pass through them. Trees may also block prevailing winds and filter dust from the air.

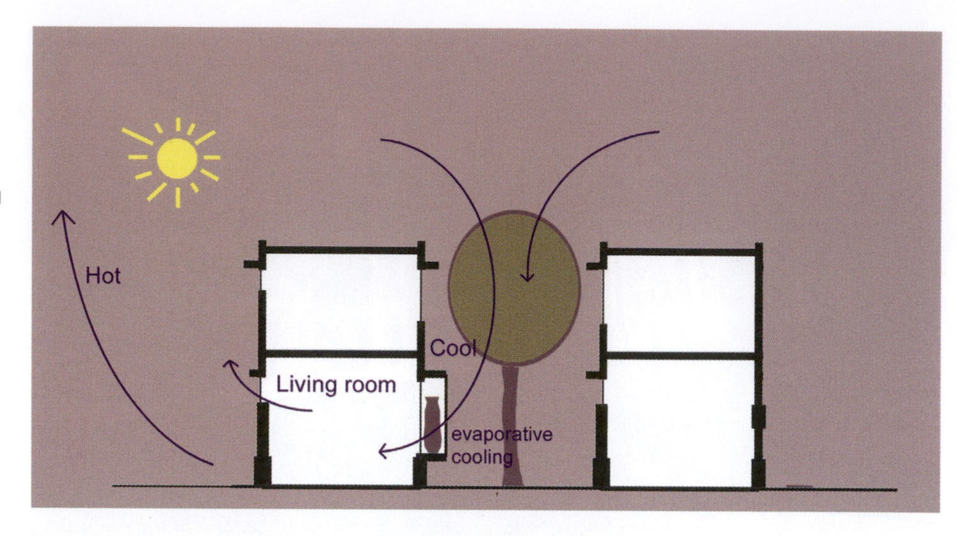

Figure 2.17 Courtyards in themselves provide shade, but adding trees provides an extra cooling effect. These buildings also use small windows on the sun-facing side with overhangs. In addition, the picture illustrates traditional evaporative cooling by drawing the air around an earthenware pot containing water.

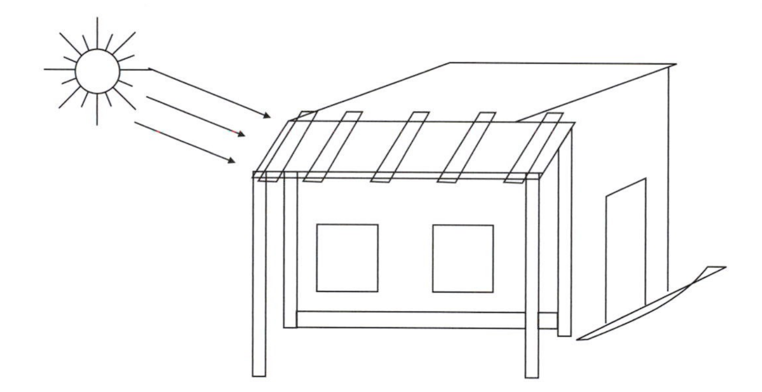

Figure 2.18 Pergola for the shading of facades.

Source: Kishore, 2009

Overhangs

Overhangs above windows are a common form of shading device. They should be sized relative to the latitude, location and window size, so that the amount of light admitted through the window at different times of the year prevents overheating in summer and helps solar gain in winter. In the example shown in Figure 2.21, the building is assumed to be at latitude 40°N.

In summer, the unwanted direct beams from the overhead noon sun are blocked from entering the building to prevent overheating, but in mid-winter, at its lowest angle, the now welcome rays are allowed to reach and warm the rear wall.

Natural ventilation

This involves utilizing the stack effect in the design of the building so that cool air enters and passes through the building, replacing or pushing out warm air, which is allowed to escape at the top of the building. Pressure differences are created in two inter-related ways: differences in temperature, and directing the wind. The cool air could come from openings positioned low down on the side of the building that doesn't ever face the sun and is in permanent shade, or from underground. If this airflow is passed over water, the evaporative cooling effect increases the benefit.

The relative size of the openings is important. If the wind is constant, the best result is obtained by having a small inlet and a large outlet. If the wind is not constant, a large inlet is preferable because it lets in a greater volume of air. If a solution cannot be designed on known local conditions, then full wall openings on both sides should be used with adjustable shutters to channel the airflow in the required directions. For example, courtyards and enclosed open spaces have been used for centuries for this purpose. During the day sunshine coming into the courtyard heats up the air which rises to escape. To replace it, cool air must be drawn in at ground level. This has to come from openings in rooms facing the courtyard, which in turn must draw warm air out of the building.

At night time, the process is reversed. The warm roof surface is cooled by convection. If the surface is sloping inwards towards the courtyard, the cooler air falls down into the courtyard and enters the building. Hot air inside the building is allowed to escape from openings in the top of the building. The larger the thermal mass of a building, the more even the room temperature.

Figure 2.19 A pergola shading a French window from too much glare.

Vented wall and/or window openings must be equipped with weather, burglar and insect-protection and can use automatically controlled flaps. If daytime solar gains are reduced to a minimum, night ventilation can perform well. The building must have good passive stack ventilation design. If free ventilation is not feasible then mechanical ventilation will be more efficient if it takes place only at night.

Figure 2.20 The Welsh Assembly Government and local authority buildings in Aberystwyth demonstrate the use of several shading strategies together with large, south-facing windows to maximize solar gain and daylighting but minimize glare and overheating. The vertical axis wind turbine supplements the electricity demand.

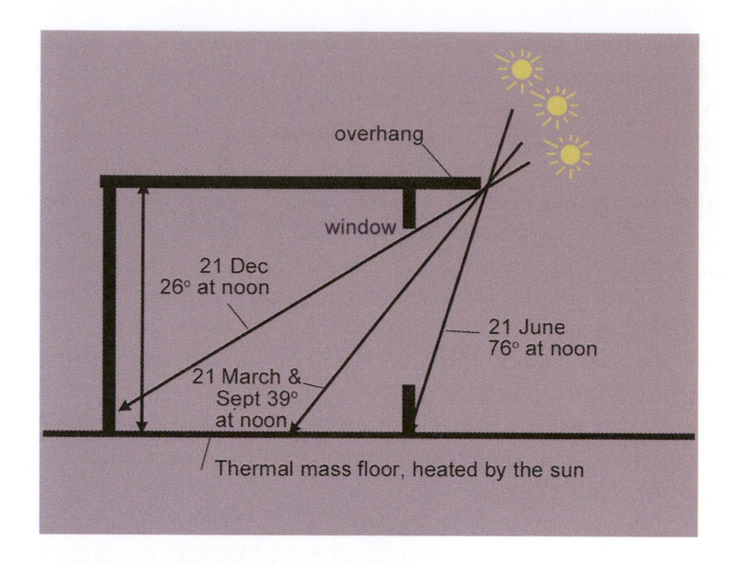

Figure 2.21 Overhang depths may be calculated according to latitude to predict where the light will fall at different times of the year and, from insolation levels, how much heat is likely to enter the building.

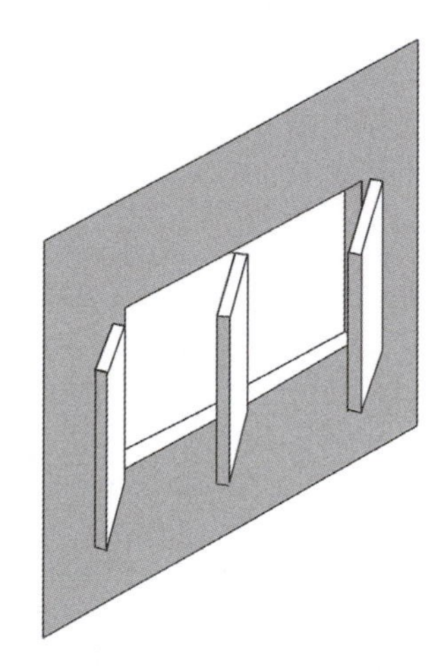

Figure 2.22 Shading can also be in the form of vertical fins, where indirect light is necessary for illumination, but where direct light would be too hot. The angle of the shading is calculated to be more or less perpendicular to the sun's path, whatever the orientation of the wall, during the hottest hours.

Figure 2.23 Shading and overhangs on the Bundespresse, Berlin, Germany.

Source: © Frank Jackson

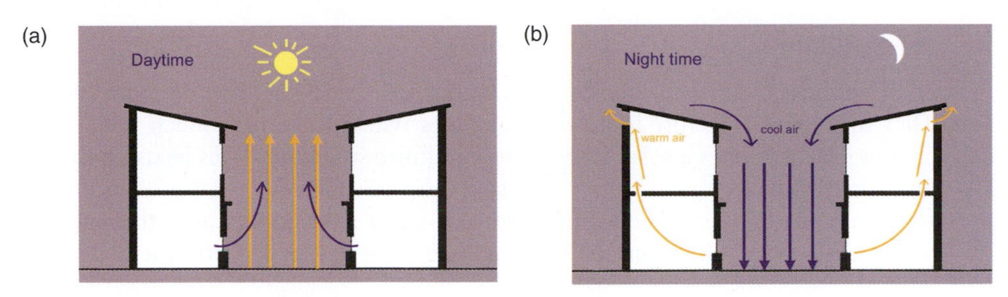

Figure 2.24 (a) In the daytime, hot air rising in a courtyard draws cooler air from inside the building. (b) During the night, cool air falling from the roof into the courtyard is drawn into the building to replace the warm air rising out of openings at the top.

Wind towers

Wind towers, used in hot and arid locations, take this principle to its logical extreme. They capture and use the local wind. The tower is located to the side of or away from the building to be ventilated. It has openings in the side at the top, designed to face the prevailing wind. Inside, it is separated into two or more shafts, which allow air to move easily up and down the tower at the same time.

In the daytime, ambient air blows into the openings. It is drawn down by pressure differences to the base, and sometimes even underground in a basement, where it cools. It is then allowed to circulate upwards through the building and exit through openings near the top. This can cause a perceptible breeze on the skin of an occupant, which is very pleasant. At night time, there is a reversal of airflow; cooler air enters the bottom of the tower after passing through the rooms. It is heated up by the warm surface of the wind tower and leaves in the reverse direction, sucking warm air out of the building. In this way, the heat in the thermal mass of the building, gained during the day, is allowed to escape.

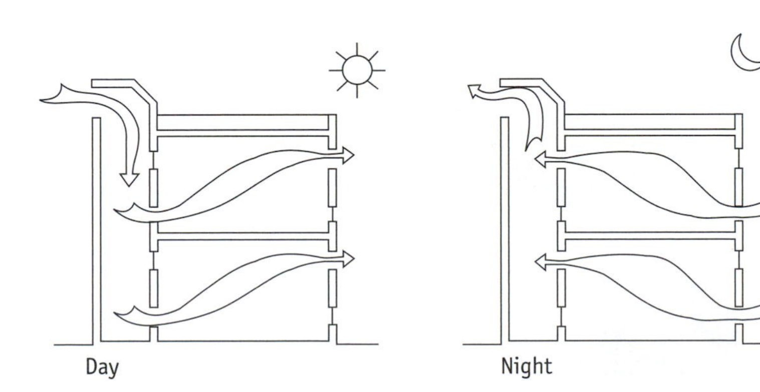

Figure 2.25 Daytime and night-time operations of a wind tower.

Source: Bansal, Minke and Hauser, 1994; Kishore, 2009

Solar chimneys

Solar chimneys use air heated by solar radiation to suck out hot air from inside. They are useful therefore in areas or at times where there isn't much wind. The chimney incorporates a solar collector to maximize solar gain. This heats up the air inside it. Temperatures can get very hot so it must be isolated by a layer of insulation from inhabited spaces. The air inside the chimney rises and escapes through the top, sucking after it air from the building below. Its power depends on:

- the size of the collector – the larger it is, the more heat is collected;
- the size of the inlet and outlet – an inverted 'funnel' ration is best, with the narrow opening at the top;
- the vertical distance between the inlet and outlet – a longer distance creates greater pressure differentials.

The collector must be on the sun-facing side of the chimney, and can be near its top or cover the whole side of the chimney. It will be painted black and glazed to absorb more heat.

Air inlets and heat exchangers

Buildings that incorporate wind towers, solar chimneys or other kinds of natural ventilation can also use an underground heat exchanger. This draws air underground from somewhere outside the building. It sucks it (either naturally

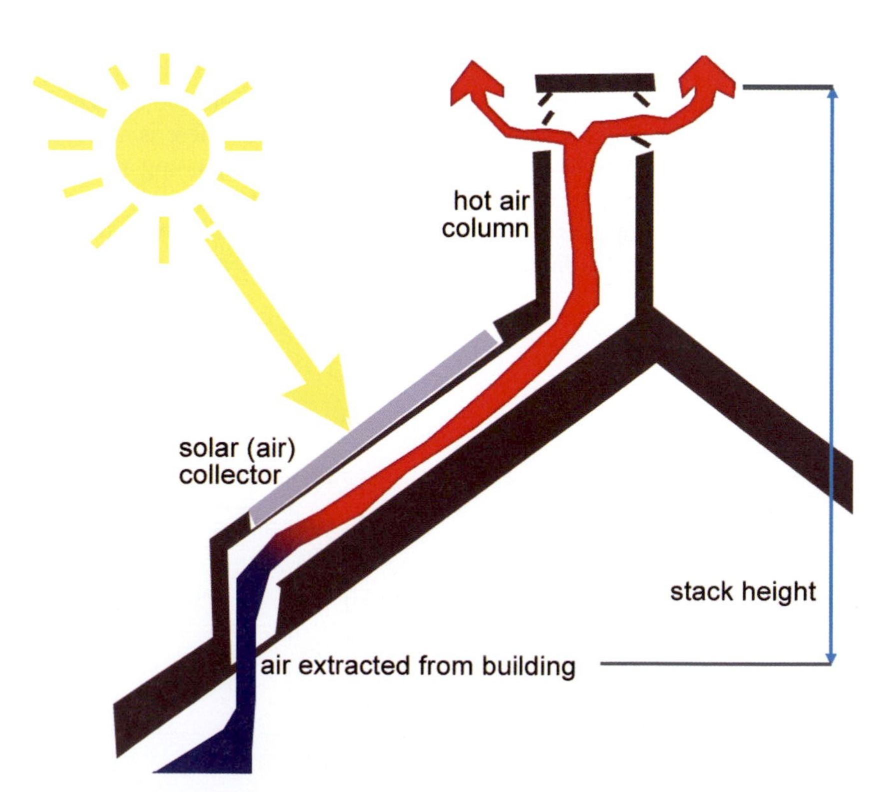

Figure 2.26 Principle of how a solar chimney works. The sun's heat warms the chimney, and the air inside, drawing cooler air from the interior.

or with a fan) through a heat exchanger to absorb the heat from the air into the ground. This cooled air then enters the building at the bottom and is sucked through it and out by the stack effect. Remember at all times that passing incoming air over a body of water will help to cool it. If the winters are cold, the same system can be used to warm the incoming air. This system is a ground source heat pump.

Evaporative cooling

When water changes from liquid to vapour, or even from solid to liquid, it absorbs heat. The change is called a phase change. Indoor pools can use this effect and be a pleasant addition to an indoor space but do require maintenance. The cooling power is dependent on the volume of evaporating water, air temperature, velocity and humidity.

Indoor plants and water

A similar effect is generated from the leaf-surfaces of indoor plants. Indoor plants are also known to remove pollutants, such as particulates, dust and unhealthy gases from the air, expelling oxygen and moderating the indoor humidity and temperature. They also buffer sound, muting ambient noise. By close planting, a cooling power of 16W/m^2 can result due to the evaporating power, according to Preisack et al (2002), who have researched the relative benefits of different plants, and found that some are far more beneficial than others.

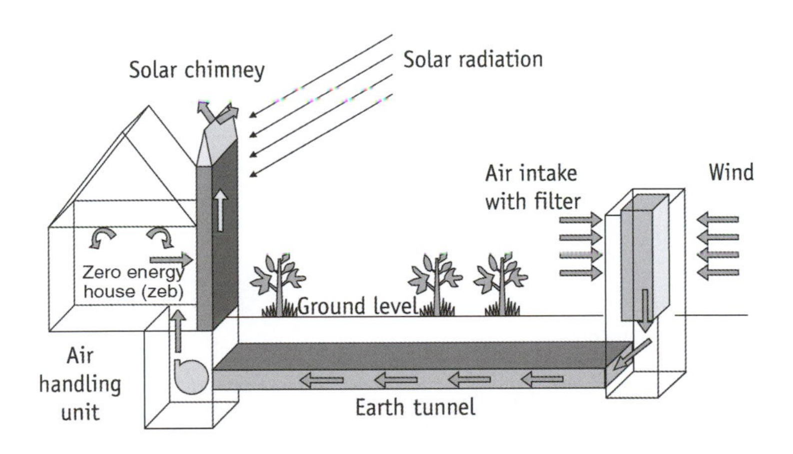

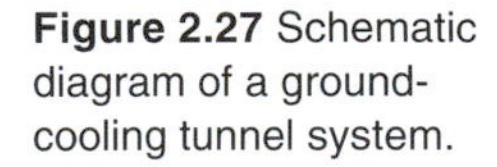

Figure 2.27 Schematic diagram of a ground-cooling tunnel system.

Source: Kishore, 2009

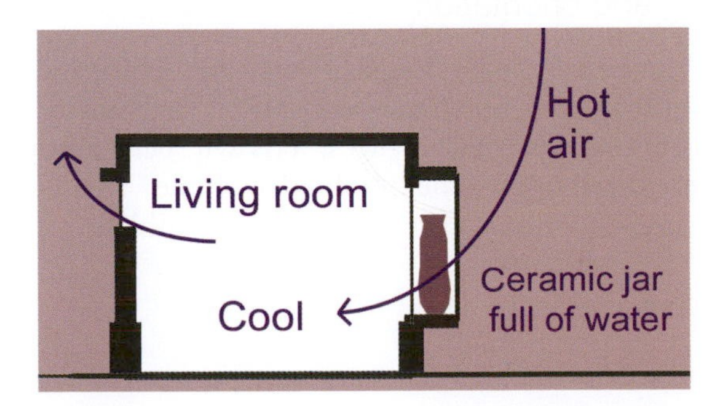

Figure 2.28 Water placed in the path of incoming hot air will draw heat from it by evaporative cooling, as it changes to water vapour. Plants fulfil a similar function.

Water is also used to cool air. In hotels, restaurants, office foyers and so on, water basins, indoor ponds, artificial creeks and landscapes, water walls and indoor fountains are often found, integrated aesthetically into the building. The slow evaporation of water removes heat from incoming breezes. The cooled air can be directed through the building using the stack effect.

Phase change materials

Other materials also undergo phase changes within the temperature range found inside buildings. An example is wax. When wax melts, it absorbs heat. When it solidifies it gives off heat. This property is used in wall coverings to modulate the temperature. Additives to paraffin wax can precisely adjust the temperature at which it moves from one phase to another to match it with the difference between day- and night-time temperatures. It can then heat up in the daytime and melt, and at night re-solidify, giving off the heat absorbed in the day. As a result, it helps to modulate the temperature over 24 hours.

Other phase-change materials (PCMs) are: salt hydrates, fatty acids and esters and various paraffins (such as octadecane). They can come encapsulated within tiny cells in paint or in wall coverings to be placed on areas of a building that receive direct solar gain. They are also used in stand-alone air conditioning units, with fans. PCMs are also very widely used in tropical regions in telecom shelters.

Cooler climates

A similar set of considerations apply to passive solar building design for cooler climates, but with contrasting aims. Again, the variables are:

- spatial planning;
- building orientation;
- thermal mass;
- window position and size;
- external wall coverings;
- shading and overhangs;
- passive ventilation;
- phase change materials.

We will consider these one by one.

Spatial planning and orientation

The best way to maximize the potential of solar gain is for the building to face the equator – south in the northern hemisphere and north in the southern hemisphere. Elongating it on the east–west axis will permit a larger amount of glazing to be used in relation to the building's volume. This heat and light can then be directed into the rooms at the back of the building.

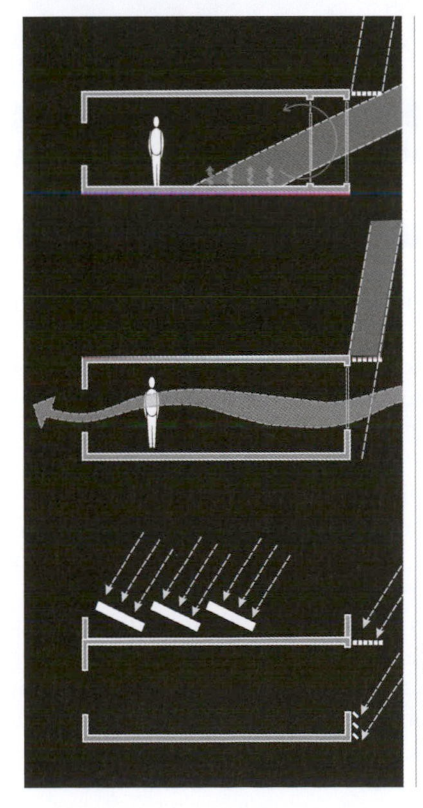

Passive solar heating: *A combination of south-facing windows, light shelves over the windows and night-time insulation can produce as much as 75 per cent of each unit's heat.*

Natural cooling: *Cross ventilation, green roofs and proper shading make air conditioning necessary only in July and August, Tianjin's hottest months, the team determined.*

Community generators: The architects suggest that heat and power requirements for each unit can be met by a combination of solar thermal and photovoltaic modules,wind turbines mounted on the roof of the taller buildings and the conversion of biodegradable waste into gas.

Figure 2.29 A proposed ultra low-impact development in Tianjin, on the east coast of China, latitude 40°N. This high-density block of accommodation would be almost completely self-sufficient. The passive solar aspects of these blocks are indicated in the diagram. The area receives between 175 and 265 hours of sunshine a month, even in July and August which is the monsoon period, totalling 2522 hours per year.

Source: Mott MacDonald

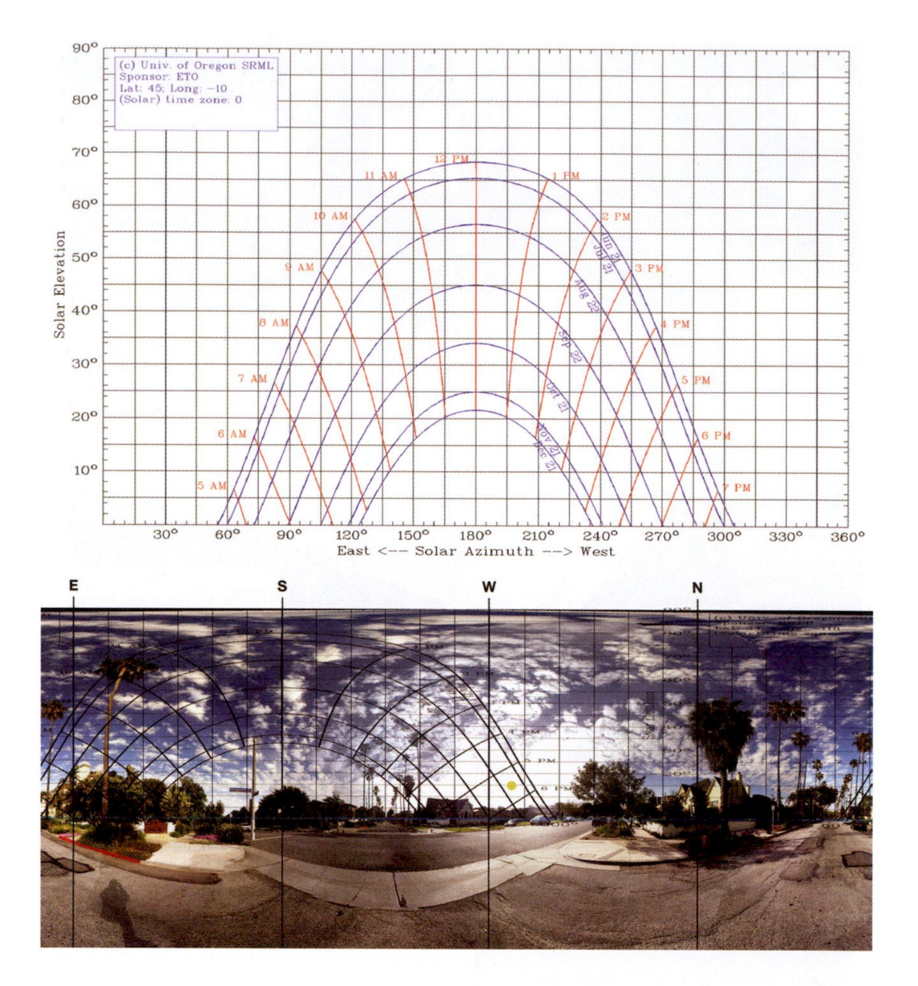

Figure 2.30 The way the time of year affects the sun's position relative to the building can be calculated by inputting the latitude and longitude of the location into a free online program at http://solardat.uoregon.edu/SunChartProgram.html. The result is a sun path diagram. This could even be superimposed using graphics software on a photograph of the location, as in the illustration, to easily visualise the sun's azimuth angle at any time of year and time of day.

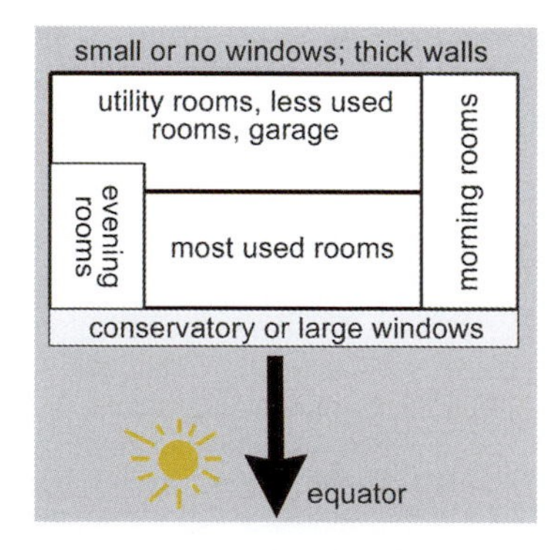

Figure 2.31 The room plan should – if it is a dwelling – incorporate the main living rooms on the equator-facing side, with utility rooms, less used rooms and a garage if any on the north side.

If it is not possible to orientate the building this way, then an orientation 20° east or west of north–south will only risk losing around 6 per cent of the maximum solar gain; 25° will lose 10 per cent. There are many permutations of interesting housing estate layouts in which every home has an equator-facing aspect but still retains privacy.

In designing a new building in the northern hemisphere, it is important to ensure that as many of the windows as possible are placed on the south elevation. East is also useful for morning rooms and west for watching sunsets. On the north side, windows should be kept to a minimum and as small as possible for lighting, because heat is more likely to escape through a window than a wall (see Figure 2.31). This wall, which receives no sunshine at all, should also have high thermal mass and/or be externally insulated, to retain heat in the building. The converse is the case south of the equator. Large windows inevitably compromise residents' privacy; but blocking light ruins the point of having them. The use of angled blinds, selective coatings or a sensitive estate plan (Figure 2.32) to avoid overlooking of windows are all ways to retain privacy.

Floors and interior walls with high thermal mass should be in line to receive the incoming light so that they can store it for later release. This means tiled solid floors and dense solid walls should not be covered by carpets, rugs or hangings.

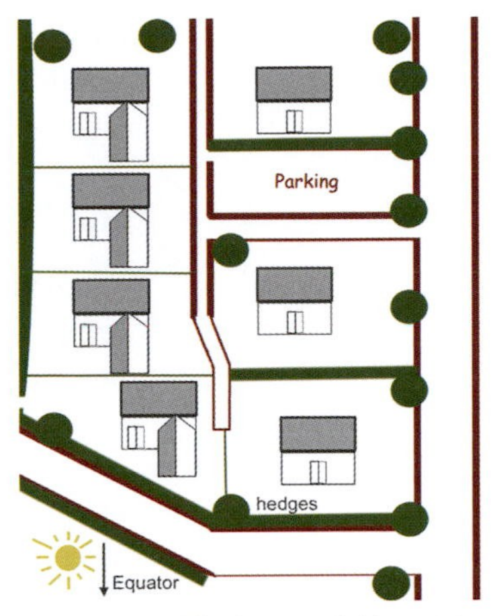

Figure 2.32 The layout of this estate allows all houses to face the equator for maximum passive solar gain, while providing privacy due to window positioning, extensions and hedges.

Source: © P. Borer

Conservatories

Conservatories can be seen as an extension of the Trombe Wall principle (see Chapter 4) – except a whole new room is now glazed, in front of the thermally massive wall. The floor of the conservatory needs also to have high thermal mass and be thoroughly insulated underneath. It helps if there are gaps at the bottom and the top of the wall adjoining the rest of the building to encourage a convection loop to form. Alternatively, a ducting system can be employed to direct the heat further into the building.

Conservatives may also be on the west side of the building, as well as the equator-facing side, where they capture the heat late in the day. They are often pleasant places to sit, but unless superinsulated, with a solid, insulated roof and high-performance glazing, must not be

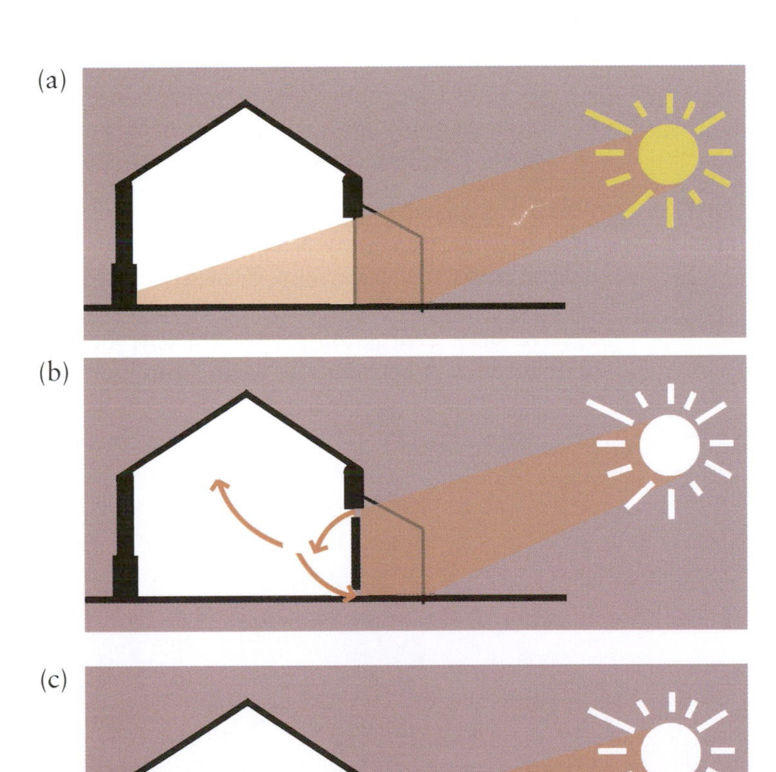

Figure 2.33 Ways in which conservatories can conduct heat into the rest of the building: directly, by convection through gaps at the top and bottom of the rear wall, and by convection plus radiation from a heated thermal mass wall. Ducting from the top gap of the rear wall could also transport the heat into rooms further inside larger buildings. Duct lengths must be as straight as possible and always rise to take advantage of the stack effect.

treated as an extra room that needs to be heated in colder weather, as this would be counter-productive. They can be added to existing solid-walled and cavity-walled structures.

Rooflights

Glazing in a roof, which sees more of the sky, can be beneficial as long as precautions are taken to avoid overheating, and to avoid heat loss at night and in cold periods. The principles are the same as above, so the sun's light should be allowed to fall onto a thermally massive surface, such as a wall.

Heat distribution and ventilation

The space that benefits from the direct solar gain obtained from the above techniques will usually be the one that gets the sunlight. But the design can allow it to be conducted, radiated or convected to other parts, using passive stack ventilation, pre-warmed air, or Mechanical Ventilation with Heat Recovery (MVHR). Adequate ventilation is necessary to remove internal pollutants and control humidity. The principle is to 'build tight – ventilate right'. The principal options are:

- a whole-house passive stack ventilation system;
- MVHR – wet rooms only or whole house;
- a whole-house passive stack ventilation system with incoming air pre-warmed underground using a ground-source heat pump.

Whole-house passive stack ventilation

Replacement air is drawn into the property through trickle ventilators in the habitable rooms, especially those (like bathrooms and kitchens) with high humidity. Interior doors must not be draught-proofed, to allow air to circulate. Passive stack ventilation saves electricity because it doesn't use a pump, but, neither does it reclaim the heat from the lost air. To minimize heat loss,

Figure 2.34 A skylight or clerestory can admit more light and heat.

Figure 2.35 Skylights, clerestories or atriums can be used to direct sunlight further into the building. An area in the centre of the building can therefore be used as a collector and absorb heat throughout the day. The heat can then be distributed around the building.

Figure 2.36 The roof of the railway station in Berlin, Germany, which uses solar glazing and automatic window openings for cooling.

Source: © Frank Jackson

humidity-sensitive duct inlet grilles should be installed to ensure that the system only works when needed. Ducting may need to be installed, leading from key rooms in the house into the loft space.

Night ventilation

Day-warmed thermal mass can be cooled by night-time passive stack ventilation through the roof, when outdoor temperatures are – at least in central Europe – mostly under 21°C (70°F) and room temperatures are higher. Incoming replacement air is drawn from ground floor inlets opened at night. These vented wall and/or window openings must be equipped with weather, burglar and insect-protection and automatically controlled flaps. The larger the thermal mass of a building, the more even the internal temperature.

Office case study – *the Solar XXI building in Lisbon*

This 1500m^2 (16,146ft^2) multipurpose building in Lisbon, Portugal, is naturally ventilated and functions as a near zero energy building. Its cost is said to be little more than a conventional building of the same size. The office space is on the south side of the building to take advantage of daylighting and solar heating. Spaces with

(a)

(b)

Figure 2.37 The south-facing (a) and north-facing (b) sides of a new Passivhaus standard dwelling in Wales, UK, a three-bed property, called 'Larch House' after its locally sourced Larch cladding. Note the large windows on the equator-facing side and small windows on the cold side. It also features closed-panel timber framing to minimize draught, superinsulation, local stone for thermal mass, and thermal and photovoltaic modules. It achieved outstanding levels of air-tightness, exceeding the Passivhaus standard of less than or equal to 0.6 air changes per hour at 50 Pascals by recording on average 0.197 air changes per hour.

Source: BRE

Box 2.2 The thermal envelope

In building parlance, the skin of a structure is called its 'thermal envelope'. In a passive solar building in cooler climates, this must be optimized to retain the solar heat you've captured. In a hot climate, a similar practice keeps out the heat and reduces cooling bills. This involves:

- a continuous air barrier;
- a completely continuous, high level of insulation – no gaps;
- high-performance windows and doors;
- minimal thermal bridges;
- preserving or enhancing breathability – the ability of the walls to absorb moisture on the inside.

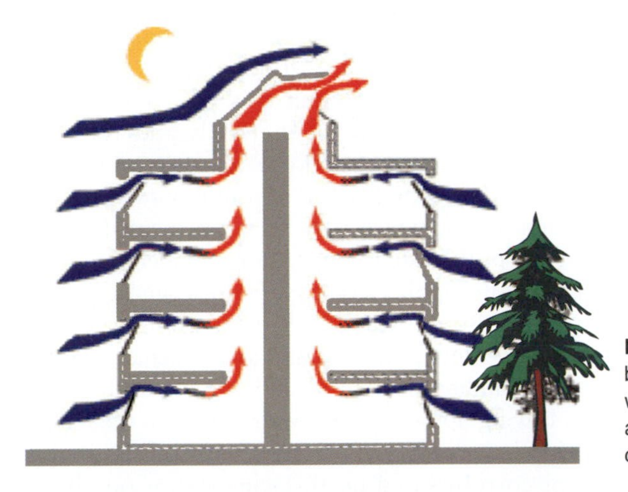

Figure 2.38 Night ventilation: flaps, behind grilles in the facade, or tilted windows, are opened automatically to allow heat to escape and be replaced by cooler night air.

Figure 2.39 The Solar XXI building in Lisbon, Portugal, which functions as a combined office and laboratory at the National Energy and Geology Laboratory (LNEG).

Source: International Energy Agency (IEA)

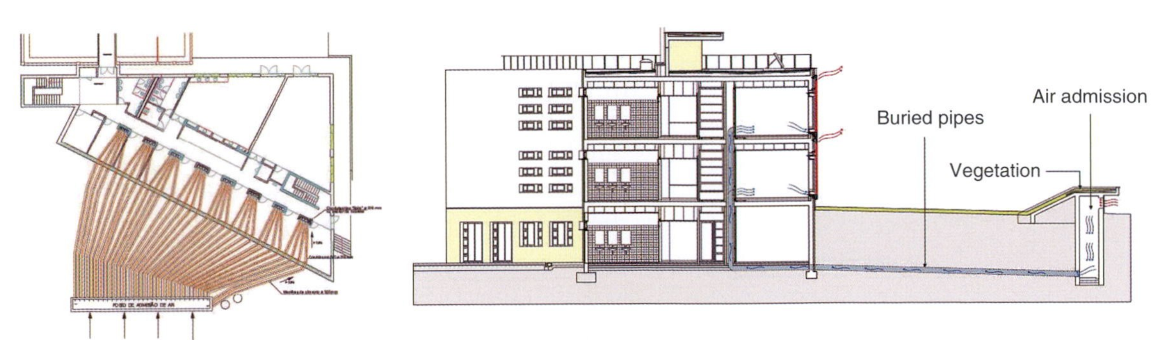

Figure 2.40 Plan and sectional view showing distribution of the buried air pre-cooling system.

Source: IEA

Figure 2.41 Compound parabolic concentrating (CPC) thermal collectors on the building roof.

Source: IEA

intermittent use, such as laboratories and meeting rooms, are on the north side of the building. Office spaces are in use from 9am to 6pm weekdays, and the ventilation pattern was arranged to suit this. The building has high thermal capacity using external installation on the walls and roof. The south facade supports 100m^2 of solar PV modules and the majority of the glazing. Additional space heating is provided by 16m^2 (172ft^2) of roof-mounted solar thermal collectors that also supply hot water, which can be supplemented by a gas boiler. The 18 kilowatt-peak (kWp; the rated power output under standard test conditions) grid-connected PV arrays supply electricity; further panels are located in a nearby car park where they also provide shade. The entire system satisfies heating requirements of 6.6kWh/m^2 and cooling requirements of 25kWh/m^2. Annual electricity use for the building is about 17kWh/m^2, of which 12kWh/m^2 is supplied by the PV arrays, leaving 30 per cent to be drawn from the national grid.

Natural lighting is encouraged – in the centre of the building is a skylight providing light for corridors and north-facing rooms on all three storeys. The installed artificial lighting load is 8W/m^2. There is no active cooling system. Venetian blinds are outside the glazing to limit direct solar gain. Natural ventilation is promoted through the use of openings in the facade and between internal spaces, together with clerestory windows at roof level, which help create a cross wind and stack effect. Assisted ventilation is provided by convection due to the PV module heat losses. To supplement this in the cooling season, incoming air can be pre-cooled by being drawn by small fans through

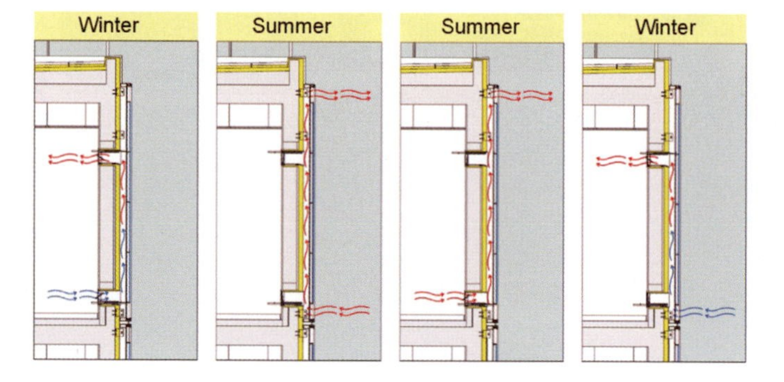

Figure 2.42 Cross and vertical ventilation systems acting together with the buried pipe system.

Source: IEA

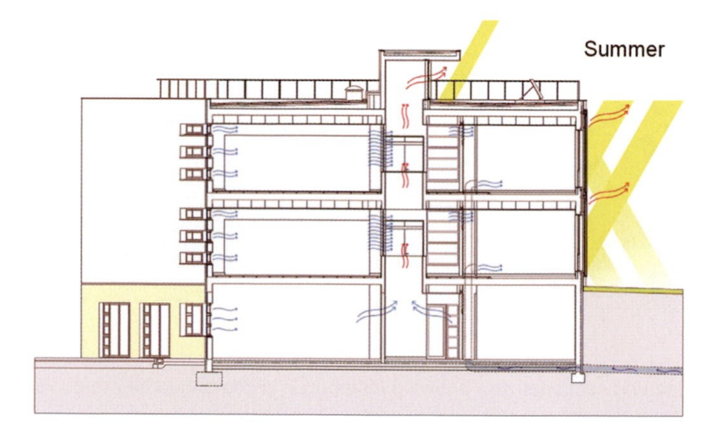

Figure 2.43 Method of operation of the heat output of the PV modules to supplement ventilation.

Source: IEA

an array of underground pipes as shown in Figures 2.40–2.43. The openings are adjustable and air is allowed to rise through the central light well. The vents are manually operable, and staff needed to be educated in their use. In other buildings, such vents can operate automatically, governed by sensors. The building occupants have been surveyed and expressed 70–95 per cent satisfaction with aspects of the air quality and temperature.

MVHR

Single room intermittent

Figures 2.44 and 2.45 illustrate a system where a room – typically a bathroom or kitchen – has outlet and inlet grilles in the outside walls, preferably some distance apart so the inlet is not sucking in expelled air. The grilles are situated for the incoming and extracted air inside the room using the same principle. Ducts from each lead to the heat exchanger unit, which may lie beneath a panel in the floor for easy access. Humidistat sensors (e.g. above the shower and toilet) operate the fans when the humidity reaches a certain level.

(a)

(b)

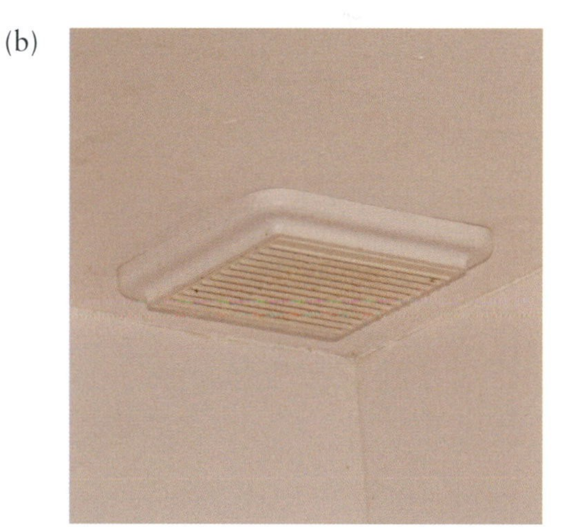

(c)

Figure 2.44 Sensor and an insulated heat exchange unit in a bathroom single-room ventilation with heat recovery system: (a) heat exchanger; (b) intake, situated above shower, removing warm, humid air; (c) output from the heat exchanger, inputting fresh air with recovered heat.

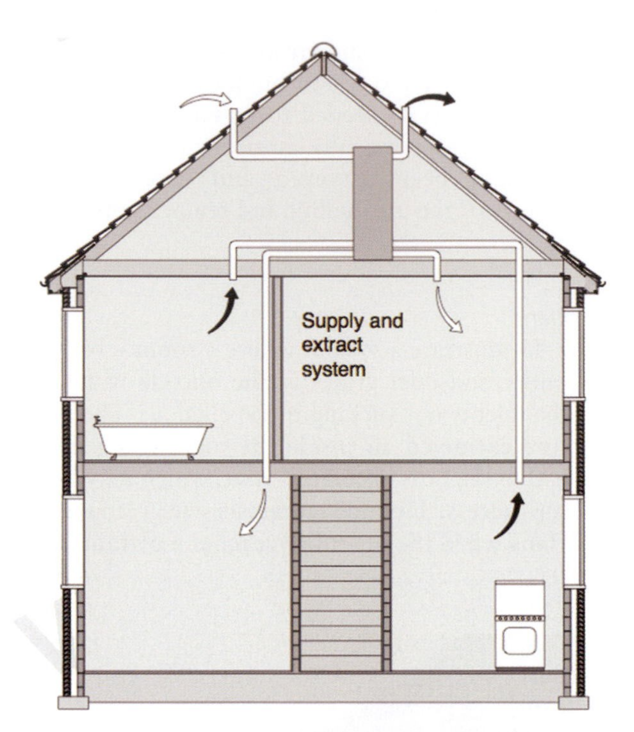

Figure 2.45 Whole-house system layout for mechanical ventilation with heat recovery.

Source: © EST

Figure 2.46 Insulated heat exchange unit in an attic – whole-house system.

Source: © Chris Twinn

Whole-house ventilation

A whole-house ventilation system may be installed if a blower test for airtightness gives a result of $5m^3/hr/m^2$ at 50 Pascals. Ducting takes air – usually from the wet rooms – to the heat exchanger in the loft, where a low-wattage pump pushes the heated, clean, incoming air down (an)other duct(s) into a lower room. Incoming air is taken from a vent in the roof. The pump can be powered by a PV module, but then will only work when the sun is shining.

Advanced systems can reverse the heat transfer effect in summer, providing cooling – air conditioning– very efficiently.

Air conditioning with a heat pump

Heat pumps take heat from the ground, air or a nearby body of water, if available. All of these systems in principle work like a fridge, but in reverse. In other words, they concentrate low-temperature heat from a larger volume into a smaller volume at a higher temperature. Ground-source heat pumps are the most efficient, and therefore provide the biggest cash and carbon savings. Typically, for every unit of energy expended on the pump, three or four units of heat energy are brought inside. As electricity is being used for this purpose, it should preferably be renewably sourced. If it is not, and the mains power is at least 70 per cent derived from burning fossil fuels, then using a modern 95 per cent efficient boiler running on a fossil fuel such as gas for heating instead would be just as efficient, due to efficiency losses on the way from the fuel in the power plant to the pump. The exception to this is if a coefficient of performance (COP) of 4.0 or above for the ground source heat pump to underfloor heating could be consistently achieved.

Figure 2.47 Trench with a coil for a ground source collector for a heat pump, prior to burial.

Source: © John Cantor

In the ventilation system illustrated in Figures 2.47 and 2.48, air is taken underground from an intake outside, and run through a coil buried outside, at least 3m (10ft) underground. The air is then drawn into the house, and let out through grills beneath the large, equator-facing windows. It then rises through the building using the stack effect. As the temperature of the air is not high (16–24°C; 61–75°F), the efficiency of the system is excellent. Depending on the outside temperature, this can have either a cooling or a heating effect, as the temperature

Figure 2.48 Air intake in a garden outside, taking air to an underground heat exchanger, and the output just inside, introducing the preheated air into the building.

Figure 2.49 Adjustable solar chimneys can be used to modulate the temperature and prevent overheating, as in the Inland Revenue building in Nottingham. The centre, completed in 1995, was a pioneering 'green' project in the UK. At night, the inherent thermal mass of the concrete is exploited and purged with fresh air to pre-cool the structure. At the corners of the buildings, the air within the glass block stair towers warms and rises on sunny days, giving extra drive to the ventilation system. Fabric umbrellas on the tops of the towers act as large dampers, lifting to exhaust hot air and closing, on cool days, to conserve heat. The office buildings were extensively prefabricated. The local bricks of the load-bearing piers were laid in a factory, around steel lifting rods, in storey height units.

underground will be more consistent throughout the year. For a better cooling effect, the system can be specified to work in reverse in the summer, removing hot air and transferring the heat to the ground. Heat pumps have a long life expectancy (typically 20–25 years for the equipment and up to 50 years for the ground coil).

Solar heat and light control

A dwelling with too much equator- or west-facing glass can result in excessive winter, spring, or autumn day heating and too much glare. Using the sunshine for day-lighting as well as heating and cooling, while avoiding too much glare, is a design challenge. Although the sun is at the same altitude six weeks before and after the solstice, the heating and cooling requirements before and after the solstice are significantly different. Modelling must account for this, using software called THERM – Two-dimensional building HEat tRansfer Modelling and Passivhaus Planning Package (PHPP) – see the Resources chapter for more information. If the building has window overhangs to provide midday shade

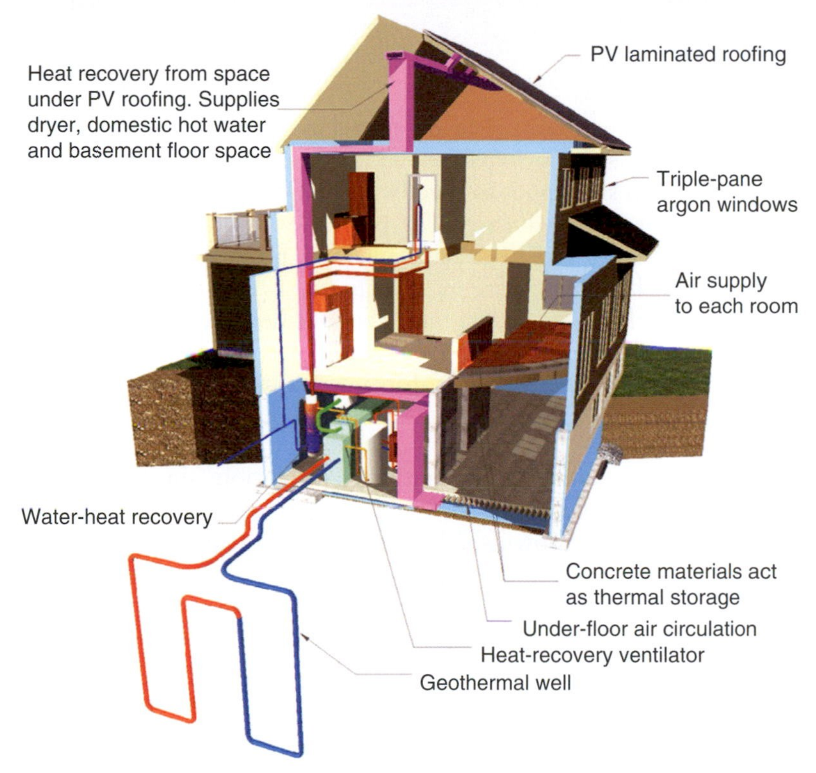

Figure 2.50 The ÉcoTerra™ house, built by Alouette Homes, is a 240m² (2600ft²) prefabricated home assembled in 2007 in Quebec, Canada. It integrates passive and active solar systems and energy-efficiency technologies. It features high-performance south-facing triple-glazed windows to maximize the amount of solar radiation entering the house while minimizing heat loss; and ventilated concrete slabs on the main and basement level to absorb incoming solar radiation and slowly release it throughout the day and into the night. Solar-heated air from a building-integrated photovoltaic and thermal (BIPV/T) system is used for space heating, domestic hot water preheating and clothes drying. Surplus electricity is exported to the national grid.

Source: IEA

Figure 2.51 External shutters, internal shutters, overhangs from balconies (visible through the window), and curtains closed in the daytime can all help keep unwanted heat out.

they should be of an appropriate depth relative to the building's latitude. However, these can still be bypassed when the sun is low in the sky. One permanent and fixed solution is to specify a 'solar control' coating on the inside of the first pane of glass, which can be precisely calibrated to reflect a given proportion of the sun's radiation back out and prevent it from entering the room. This might be appropriate in an equator- or west-facing conservatory, which tends to overheat. Such glass might be installed in a roof and reflect, say, 70 per cent of the heat.

Variable solutions are available for occupants to adjust according to conditions, such as:

- inside: window quilts, bi-fold interior insulation shutters, manual or motorized interior insulated drapes and shutters;
- outside: shutters, roll-down shade screens or retractable awnings.

Shading devices have been used for – in some cases – centuries in Mediterranean countries. They can help control daily/hourly variations. The adoption of different types of shading devices can reduce the energy requirement for cooling from 50 per cent to 100 per cent. Automated systems are on the market that monitor temperature, sunlight, time of day and room occupancy and control motorized window-shading-and-insulation devices. While it is important to take care that the energy cost of these, including the embodied energy of manufacture and installation, is not greater than that saved by reducing the cooling demand, research on buildings in Europe has discovered that the adoption of such automatic systems allows much greater solar protection and correspondingly greater reduction of energy needs for cooling. It is advantageous for the designer to consider that combining solar protection, ventilation and thermal insulation in one building component, such as a window or facade, in new-build or refurbishment, offers the potential for great savings.

Windows that receive too much summer sun but need the winter sun can also be shaded by deciduous trees. In the summer they block light, and in the winter they let it in.

Windows

As a rough guide, window area should be equal to no more than a quarter of a home's floor area, and at least half should be located on the equator-facing side, if they are not shaded. In winter, windows lose much more heat than walls, and this is especially noticeable in conservatories.

Box 2.3 Erlenweg Apartment Building in Volketswil, Switzerland

Not all solar-designed accommodation needs to shout about its solar features. In fact many people do not want to live in what looks ostensibly like an eco-home. This apartment building, located in the suburbs of Zurich, Switzerland, was modestly renovated to achieve higher energy efficiency, improve the facade's insulation, add solar evacuated-tube collectors on the roof and a ventilation system while meeting the owner's budget.

The renovation included an insulated facade, roof, basement ceilings, and heat and hot water pipes. Other solar features added were new sun blinds on the south facade, insulated shutter boxes all around the building and a new glazed entrance. The reduction in energy use was 62 per cent.

Figure 2.52 Erlenweg Apartment Building.

Source: © IEA-SHC (2009-11-SolarUpdate)

Box 2.4 U-values and R-values

U-values and R-values are measures of how fast heat passes through a material or building element, such as a window. They tell you how good an insulator it is. In the US, R-values are used. Almost everywhere else, they are referred to as U-values.

U-values are British thermal units (Btu) per hour per square foot per degree Fahrenheit (imperial); or watts per hour per square metre per degree Celsius. The lower the U-value, the better; R-values are the inverse, therefore the higher the value the better: m^2°C·h/W (metric), or ft^2°F·h/Btu (imperial). R-values are frequently cited without units, e.g. R-3.5. One R-value (metric) is equivalent to 5.67446 R-value (US) or one R-value (US) is equivalent to 0.1761 R-value (metric). Usually, the appropriate units can be inferred from the context and their magnitudes.

Modern windows are rated by national bodies and come with a declaratory label. The label should cover the U-value of the whole window including the frame. In temperate and colder climates, choose windows with U-values of 0.35Btu/hr × ft^2 × °F or below. Some triple-pane windows have U-values as low as 0.15Btu/hr × ft^2 × °F. In semi-tropical and tropical climates, windows should have U-values of 0.60Btu/hr × ft^2 × °F or less. For comparison, single-pane windows with clear glass have a U-value of about 1.0Btu/hr × ft^2 × °F depending on the frame material. These are US/imperial values. The SI equivalents would be over five times larger than the equivalent US one. For example, the SI Passivhaus standard for windows is a U-value not exceeding 0.80W/m^{2}K for both glazing and frames, which implies that the window frame incorporates insulation and the glazing is triple. It compares to a typical new-build practice of 1.8–2.2W/m^{2}K.

A window label commonly displays the following information:

- the rating level: A, B, C, etc.;
- the energy rating, e.g. –3kWh/(m^{2}K)/yr (= a loss of three kilowatt-hours per square metre per year);
- the U-value, e.g. 1.4W/(m^{2}K);
- the effective heat loss due to air penetration as L, e.g. 0.01W/(m^{2}K);
- the solar heat gain SHGC or G-value, e.g. 0.43.

Shading coefficients

The 'shading coefficient' quantifies the amount of solar energy transmitted through windows. In Europe, this coefficient is called the 'G-value' while in North America it is referred to as 'solar heat gain coefficient' (SHGC). G-values and SHGC values range from 0 to 1; a lower value represents less solar gain. Shading coefficient values are the sum of the primary solar transmittance

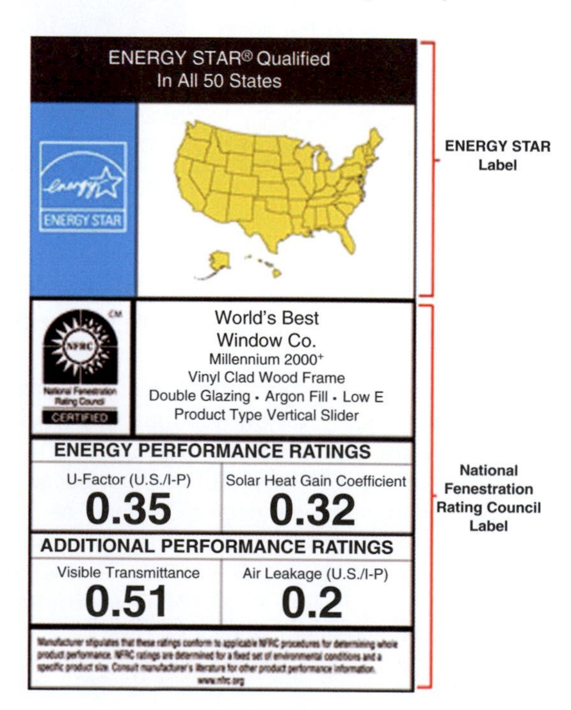

Figure 2.53 A sample American window efficiency rating label.

(T-value) – the proportion of the total solar insolation entering through the glazing, plus the secondary transmittance – the proportion absorbed in the window (or shading device).

For south-facing windows in temperate or cooler climates, an SHGC of 0.76 or greater is recommended. This can be achieved with hard-coat low-E or clear glazing. Large west-facing windows should be coated with low-E to an SHGC rating less than 0.6 to prevent overheating. In hot climates where cooling loads dominate, windows with a SHGC of 0.5 or below are preferred. Manufacturers supply a huge range of coatings: some can permit only 6 per cent of light to enter the building, or 8 per cent of heat. Individual panes specifically designed for eco-homes in higher latitudes have extra-clear outer layers, letting up to 80 per cent of light and 71 per cent of the sun's heat in, although the sum of the three panes in a low-emittance coated triple-glazed unit would reduce this to about 65 per cent.

Low-E coatings

A low-E coating on the inside pane allows short wavelength radiation inside through the pane, but longer wavelength (warm, infrared) radiation is reflected back into the building. This helps to retain the solar thermal heat within the building. ('Emittance' refers to the ability of a material's surface to emit radiant energy.)

For a quick and easy solution in retrofitting on existing glazing, fixing an adhesive window-foil to the inside of the glass can reduce heat transmittance and glare. This plastic foil lets in a certain amount of light but blocks thermal radiation above about 22°C (72°F).

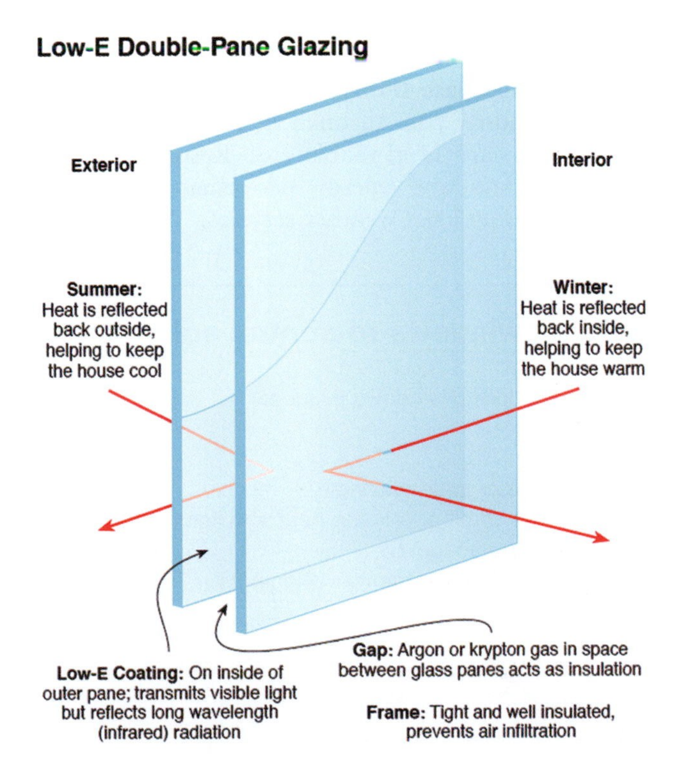

Figure 2.54 How low-E coating works on a double-glazed unit.

Figure 2.55 Triple-glazed window with insulation inside the frame to counteract thermal bridging.

Source: NorDan

Using natural light

The passive solar home will try to minimize the use of artificial lighting and its energy burden by using natural light. A number of strategies can help reduce lighting costs by 30–50 per cent while avoiding the risk of overheating.

Windows admit heat and daylight differently at different times of the year and day. This partly depends on the angle of incidence – the angle at which the sunlight strikes the surface of the glass. If this is within 20° of the perpendicular (straight on), the light will mostly pass through it; at over 35°, most of the energy will bounce off. Software is available (see Resources) to calculate cooling-and-heating degree days and energy performance, using regional climatic conditions available from local weather services.

Box 2.5 Using windows to control solar gain

Options to consider when choosing windows include:

- thickness;
- number of panes – double or triple glazing?
- coatings on the glass – to admit or release heat and light;
- cavity size or fill – argon or vacuum?
- nature of the spacer bar – to avoid thermal bridging;
- sealant used – how long will it last?
- frame type;
- frame materials – timber, ideally, with insulating thermal break;
- fixing method.

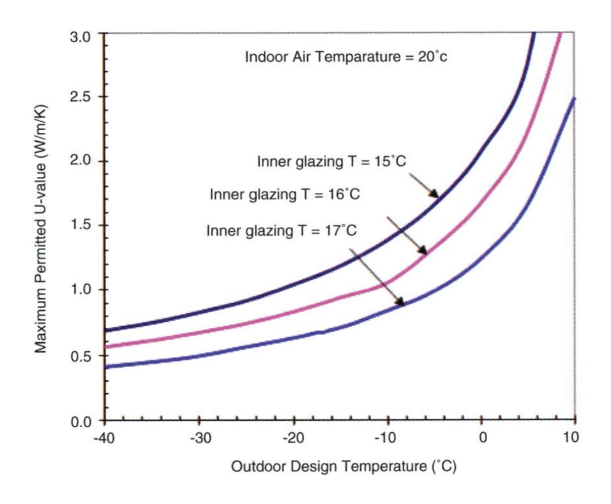

Figure 2.56 Window U-value at which perimeter heating units can be eliminated as a function of the outdoor design temperature and the coldest permitted inner glazing surface temperature.

Source: Harvey, 2008

Box 2.6 Terraced apartments in Frankfurt, Germany: Renovation to near-Passivhaus standard

These post-war buildings containing 60 apartments were renovated between 2006 and 2008. The design was worked out using PHPP software, and the result monitored for two years afterwards and found to have almost met the Passivhaus Standard by achieving an annual heat energy demand of 17kWh/(m²a). The work included 260mm of exterior insulation, triple glazing, central ventilation with heat recovery, reduction of thermal bridges, and 7.5m² (80ft²) of solar collectors on the sloped roofs of six staircases leading to the new attic apartments. A pressure test prior to the refurbishment indicated an average air change rate of 4.41/h. After refurbishment, the average was reduced to only 0.461/h. The result is dramatically shown in the thermographic images taken before and after the work.

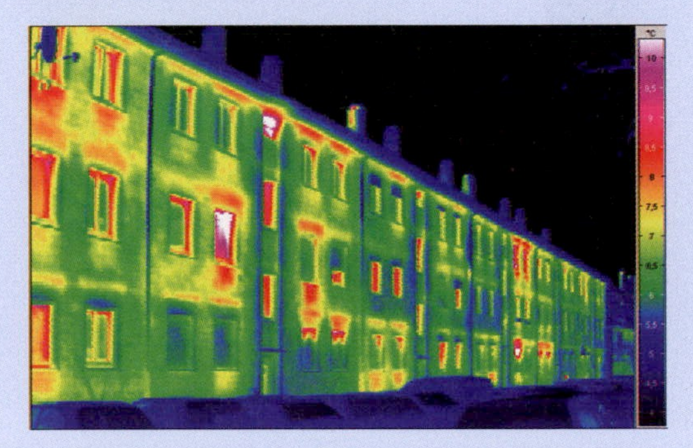

Figure 2.57 Terraced apartments in Frankfurt, Germany.

Source: © IEA-SHC (Task37-540-Frankfurt)

If a property is being built from scratch, it is possible to experiment with the window size and positioning in the design software (THERM and PHPP) to optimize the light and heat input. If the property is being renovated, then the windows are already in a fixed position. However, there are ways to improve the amount of light available from an existing window, such as:

- by painting the reveal, lintel, window sill and opposite walls a light colour to reflect more light;
- possibly slanting the sides of the reveal to increase the amount of light entering the room (Figure 2.59);
- by positioning mirrors in the reveal and opposite the window to reflect more light into the room.

Light shelves

In some places (e.g. high windows in communal areas in offices, blocks of flats and modern dwellings), installing light shelves may help to save electricity used for lighting. Light shelves reflect daylight deep into a room from horizontal overhangs above eye-level. Highly reflective surfaces bounce light onto the ceiling and up to four times the distance between the floor and the top of the window into a room. They are generally used in Continental, not tropical or desert climates, due to the intense heat gain (Figure 2.60).

Sun pipes

Light tubes or sun pipes can transport natural daylight from roofs to rooms that do not have direct access to good natural light (Figure 2.61). They are common in dense, urban areas in Japan, for example. Compared to conventional skylights, they offer better heat insulation, and are silvered inside to reflect light down them. They can help to remedy seasonal affective disorder.

Figure 2.58 Classical mastery of daylighting: the Pantheon in Rome, Italy. The building dates from around 200AD.

Source: © Clare Maynard

Figure 2.59 Angled reveal on a skylight to admit more light into the room (the white paint also helps to reflect more light inside).

Fitting

As they penetrate the airtightness barrier, doors, windows and light tubes must be installed in a way that avoids gaps. The continuity of the 'thermal envelope' of the building must be sacrosanct. Perhaps it feels counter-intuitive, but research has firmly established that even tiny gaps can have a large effect. This is because air pressures inside and outside a building can vary enormously, and a 1mm gap that is 2m long is effectively a hole $0.2m^2$ ($2.15ft^2$), which can therefore let in a lot of air.

Frames, lintels and sills must be insulated inside to prevent thermal bridging. Detailing is available online for free, for example at the UK Energy Saving Trust's Enhanced Construction Details (ECDs) minisite (see the Resources chapter).

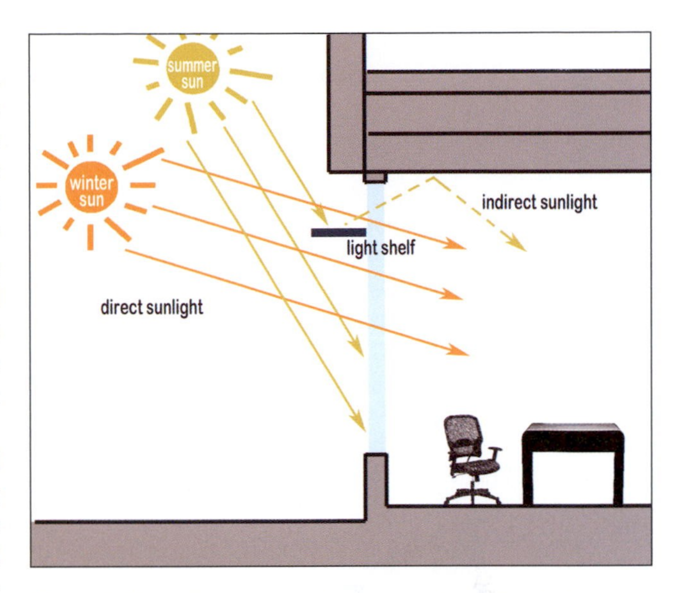

Figure 2.60 Light shelves can be positioned to reflect indirect summer light deeper into the room while blocking direct glare.

(a)

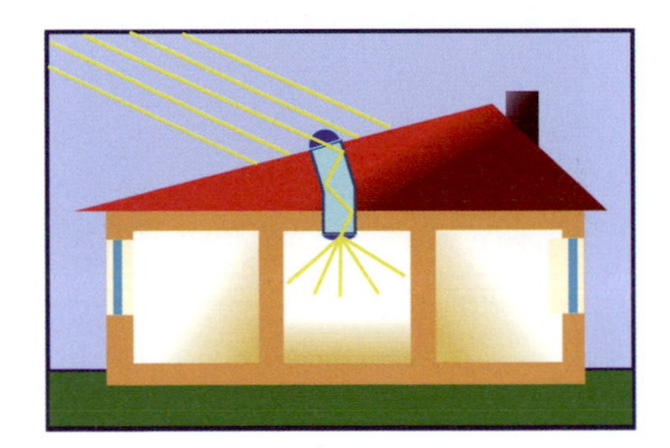

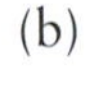
(b)

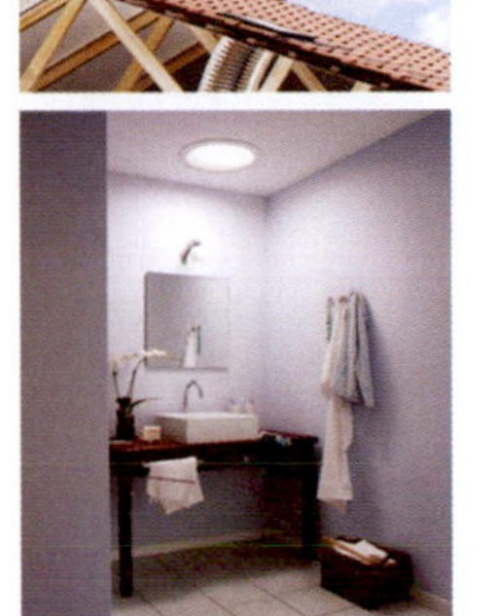

Figure 2.61 A sun tube or pipe fitted into a roof, which takes light down to a windowless bathroom.

Source: (a) German Wikipedia, http://de.wikipedia.org (b) Sun tunnel image courtesy of Greenworks and Velux

Figure 2.62 Shafts with reflective sides, opening in a ceiling, channel daylight from rooflights deep into this building.

Box 2.7 Apartment block in Linz, Austria

The apartment block shown in Figure 2.63 is located in Linz, Austria, and is almost 50 years old but in relatively good condition. The pictures show it before and after renovation in 2006, which made it exceed the Passivhaus Standard for energy efficiency by achieving an annual heat energy demand of only 11kWh/(m^{2}a). Part of the reason for its astonishing success is due to an innovative cladding system, which the designers call the 'Gap-Solar Façade'. It consists of a special cellulose honeycomb protected behind a glass facade. This is warmed by solar radiation creating a warm buffer zone in the honeycomb on the outside wall, which reduces heat losses from the interior. Efficiency depends on the amount of sunlight and the facade orientation. On the south-facing side, the losses over the heating season are drastically reduced, with an average dynamic U-value for the wall of approximately 0.08W/m^{2}K. Other features of note are triple-glazed windows, including an anti-glare shield, decentralized mechanical ventilation with heat recovery and air heating, closing off the balcony openings with windows, high insulation and domestic hot water delivered by a district heating system.

(a)

(b)

Figure 2.63 Apartment block in Linz, Austria, (a) before and (b) after renovating, including use of innovative solar cladding.

Source: © IEA-SHC

Reference

Preisack, E. B., Holzer, P. and Rodleitner, H. (2002) 'Raum-klimatisierung mit Hilfe von Pflanzen', Programmlinie Haus der Zukunft des bm:vit, Neubau Biohof Achleitner, Gebäude aus Stroh & Lehm, p42

3

Solar Water Heating

There is a city in China with 2.5 million solar water heating (SWH) systems. The city hall has passed a law obliging anyone constructing a new building less than 12 storeys high to install solar water heating unless it has a good reason not to. The city is Kunming, capital of southwestern China's Yunnan Province. Perhaps construction companies in Kunming will complain at the cost of having to install these systems. Yet at just 3000 Renminbi, it represents only one half of a per cent of the total cost of the average house. It will pay for itself within three or four years.

Kunming is not alone in China. Rizhao – it means 'sunshine' in Chinese – has over half a million square metres of solar water heating collectors with 99 per cent of households in the central districts using solar water heaters and more than 30 per cent doing so in the outlying villages.

On the other side of the world in sunny California, families in solar-panel equipped homes are also enjoying free hot water. They have solar collectors to heat their swimming pools and pump water to flat collectors on the roof. The water is pushed through tiny tubes. 'It's almost the equivalent of the garden hose sitting in your yard. Turn on the faucet and hot water comes out', solar expert Graham Owen of GoSolar Company explains. The California Energy Commission says 28,000 homes in the state are now solar and more families are installing it (NBC, 2010). According to the California Energy Commission and California Public Utilities Commission, Californian systems cost around $5000.

Most people usually associate solar panels with shiny electricity-generating modules. However, solar water heating is usually far more efficient and worthwhile an investment at a small to medium scale. The paybacks are quick. It is a well proven technology. In fact, solar thermal for heating and cooling has been the more affordable technology for the domestic and small business sectors that have been its majority adopters. Worldwide, it dominates the solar renewables market, accounting for 84 per cent, as opposed to solar PV's 14 per cent, of installed power. In 2007, over 90,000 GWh of thermal energy created by over 235 million square metres of collector area deployed across the world saved the emission into the atmosphere of over 40 million tonnes of CO_2.[1]

In the US in 2008, there were 485MWth of domestic SWH systems and 7000MWth of solar pool heating (SPH) systems. Approximately 37 per cent of all SWH systems installed were in Hawaii, followed by 20 per cent in Florida, 7 per cent in California and even smaller quantities in the other states. With over 80 million detached single-family homes in the US alone, the untapped potential for SWH as well as space heating and cooling is huge.

Figure 3.1 Kunming City in China, where solar water heating is mandatory on most buildings.

Source: Alex Wang

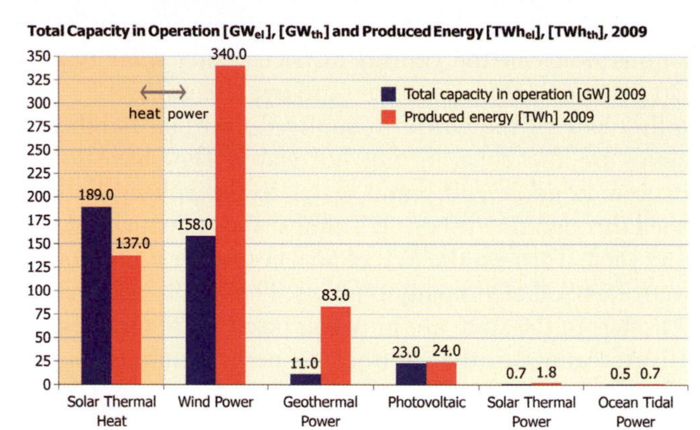

Figure 3.2 Solar heating ranks ahead of wind power in meeting global energy demand, after wood-burning and hydroelectric power, when compared to other forms of renewable energy. It yields much more than PV. Renewable energy policies still do not recognize this.

Source: © IEA

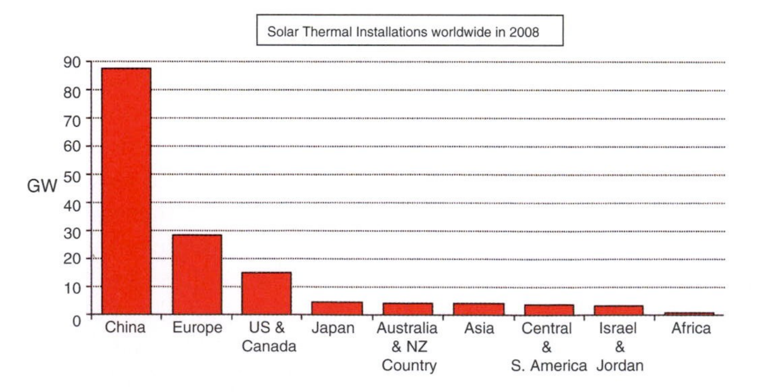

Figure 3.3 151.7GW_{th} of solar thermal collector capacity was in operation worldwide by the end of 2008, corresponding to 217 million m^2 of collection surface. Of this, 131.8GW_{th} were flat-plate and evacuated tube collectors and 18.9GW_{th} were unglazed plastic collectors. 1.2GW_{th} of air collector capacity was installed. China and Europe accounted for 86.3% of the total installations. The annual growth rate for installations since 2000 has been 20%.

A history of solar water heating

SWH is one of the oldest uses of solar power and the simplest to install. The ability of glass to trap heat was first noticed in the eighteenth century when Horace de Saussure (Figure 3.4), a Swiss naturalist, wrote in 1764, 'It is a known fact, and a fact that has probably been known for a long time, that a room, a carriage, or any other place is hotter when the rays of the sun pass through glass'. To measure this effect, de Saussure built a rectangular box out of half-inch pine, insulated the inside, and had the top covered with glass. He placed two smaller boxes inside. When exposed to the sun, the bottom box heated to 109°C (228°F).

Figure 3.4 Horace de Saussure, who invented the first solar oven in 1767. It was a well-insulated box with three layers of glass to trap thermal radiation that reached a maximum of 109°C (228°F).

Source: Wikimedia Commons

In nineteenth-century America, prospectors and farmers used a black metal tank to heat water during the day to save fuel. Then, in 1891, Clarence Kemp patented the world's first solar water heater, the Climax, in Baltimore, Maryland. By 1900, 1600 of these covered the roofs of homes in southern California, including one-third of those in Pasadena. The Climax was essentially a batch heater (see below), which combines the tank and heater. But by 1911, it was already obsolete; many new patents had been issued, including one for the Day and Night heater. William J. Bailey patented his Day and Night in 1909. He cannily separated the solar water heater into two parts: a heating element exposed to the sun, and an insulated storage unit tucked away in the house where it was warmer, so families could have sun-heated water at night and early the next morning. Just like modern collectors, the heating element consisted of pipes attached to a black-painted metal sheet placed in a glass-covered box. The use of narrow pipes speeded up the heating effect. In the face of this technological advancement, Kemp went out of business. By 1918, Bailey had sold over 4000 heaters. By 1941, half of Florida was using solar power to heat water. Why did this trend not continue? Again, we have the fossil fuel industry to thank: it was a campaign to undercut solar prices by utility company Pacific Gas and Electric that put an end to this growth and stopped solar power in the US in its tracks. The company supplied grid-connected electric water heating, with its promise of greater flexibility.

It was Israel that continued the evolution of solar water heating. In the 1950s, after the formation of the Israeli state, it experienced a fuel shortage and water heating was not allowed between 10pm and 6am. Engineer and entrepreneur Levi Yissar designed a new solar water heater and began marketing it through his company NerYah in 1953. By 1967, the year of the Six-Day War, 5 per cent of the population were using Yissar's system (Figure 3.5).

But then, and not for the first or last time, conflict fuelled by the demand for oil in the Middle East worked against the interests of solar power. Cheap oil from Iran, and from oil fields that Israel had captured during this war, made Israeli electricity cheaper and the demand for solar heaters dropped. It took the oil shortage crisis of the 1970s to provoke the passing of a law that required the installation of solar water heaters in all new homes (except towers, whose roofs were not big enough). Today, 85 per cent of the country's households use solar thermal systems. This is estimated to save 2 million barrels of oil a year – Israel has one of the highest per capita uses of solar energy in the world at

Figure 3.5 Solar collectors on roofs in Israel.

Source: Wikimedia Commons

3 per cent of the primary national energy consumption. On the nearby island of Cyprus, a large number of heaters were also installed.

Now, of course, there are examples of the SWH technology almost everywhere, establishing its credentials in most climates. In Germany, over 5 per cent of homes use SWH and over a million systems are installed. In the UK and elsewhere in the European Union with a similar climate, solar water heating has so far been used mostly to heat domestic water supplies in detached and semi-detached houses. But now effort is being put into introducing the technology into apartment blocks, hospitals, hotels and the commercial sector, just as it is in the Chinese cities of Kunming and Rizhao, and in California.

How it works

The technology is simplicity itself. Like a car left in the sun, the air heats dramatically, hindered from escaping by the glass. Furthermore, a black matt backing surface will absorb far more of the thermal radiation than a shiny white one. So any liquid passed through narrow black pipes beneath glass that is pointed at the sun will quickly heat up. It will retain more of its heat if the box and the pipes are well insulated – as are the pipes leading from the collector to the water storage tank.

What is heat and how do we collect it?

Heat is a measure of the amount of thermal energy an object contains. It is a product of temperature and mass. So a large mass contains more heat than a small mass at the same temperature – even if they occupy the same volume (because the more massive object will be denser). '*Heat gain*' is a phrase that refers to the heat gathered from the sun that is trapped using the 'greenhouse effect'. It is the ability

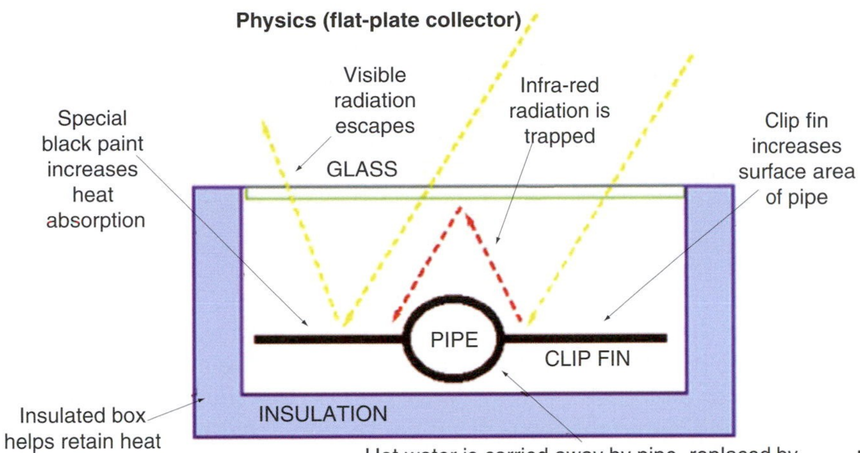

Figure 3.6 How a flat-plate solar collector collects heat.

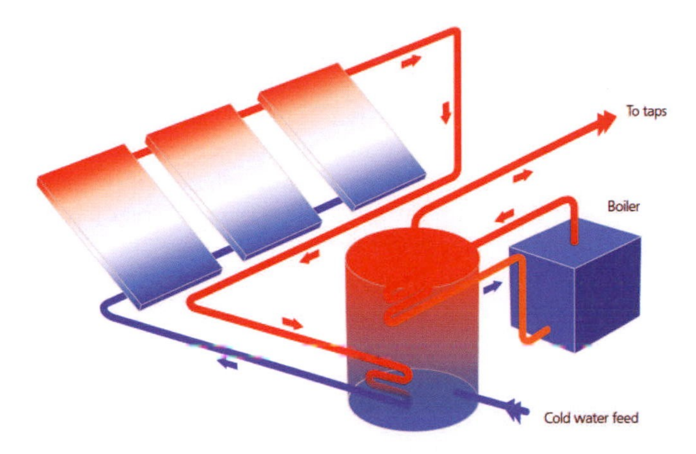

Figure 3.7 Schematic diagram of a basic closed loop solar water heating system. A pipe loop conveys heat from the collectors to the storage cylinder, where a heat exchanger transfers the heat to the water in the tank, and the liquid in the loop returns to be reheated. A second coil higher up in the tank can top up the temperature if required using another heat source, such as a gas furnace or wood stove. Water is then drawn off at the top for consumption or space heating.

Source: Energy Saving Trust

Figure 3.8 Cutaway through a flat-plate solar collector showing the insulation behind the black collector.

Source: Frank Jackson

of a glazing surface to transmit short wave radiation and reflect long wave radiation. Heat is produced when short wave radiation from the sun strikes the blackened absorber plate of the collector and becomes trapped inside it. The greenhouse effect is enhanced by special 'low-emissivity' ('low-E') coatings on the glass that prevent the long wave radiation being re-emitted from behind the glass. The best coating for a solar absorber is known as a 'selective surface' or 'black chrome' or described as Thickness Insensitive Spectrally Selective (TISS). It will absorb the most infrared radiation and emit (lose) the least.

How is that heat transferred to the water in the tank to be used? By conduction and convection. In a pan of water set to boil, kinetic energy is transferred from the fire beneath the pan by conduction to water molecules at the bottom of the pan. It causes them to move about very fast and conduct their energy to adjacent molecules. As they do so, they occupy more space than the cold molecules and rise to the top – this is convection. Cold water will sink to the bottom to replace the rising hot water, which is heated up in turn. Eventually the whole pan of water will be heated to boiling point.

In a solar collector, heat is transferred from the collector's absorber plates to the fluid inside the absorber by conduction. In a direct system, the hot water in the collector is circulated to the tank (where the hottest water will rise to the top) and comes out of the hot tap when it is opened. In an indirect system, the collector fluid is circulated through a heat exchanger coil in the bottom of the tank. The hot coil transfers its heat by conduction to the water in the tank. The hot water rises by convection to the top of the tank ready to be drawn off into the domestic heating or hot water system.

In colder climates, it is vital to separate the heat collection area from the heat storage area. If there is a risk of freezing, closed loop systems filled with propylene glycol are essential. Here, a pump circulates the heat from the collectors to the tank(s). In warmer climates, you can use simple batch heaters: water tanks enclosed inside glazed boxes.

Types of solar collectors

Collectors may be flat, batch systems or evacuated tube collectors.

Flat collectors

Flat collectors are by far the most common type of collector. In essence, they are dark-coloured absorbers through which water flows in pipes, highly insulated on every side except for the front, which is covered with low-E glass (see above). Cool water flows in at the bottom and heated water flows out at the top. Several flat-plate collectors can be linked together – either in series or in parallel. They come in grades of sophistication, which can be simply classified as glazed or unglazed.

Basic, unglazed

The basic unglazed solar collector snakes the narrow pipework – or 'solar absorber' – in a layout that maximizes exposure to the sun for a given area. Unglazed panels are mainly only found in the US and Australia.

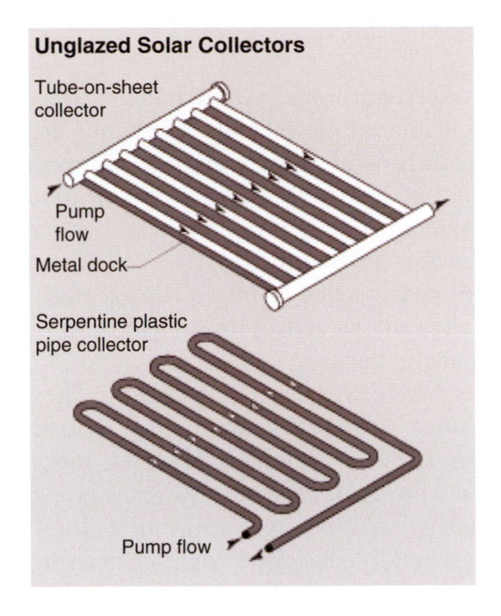

Figure 3.9 An unglazed solar collector typically used for swimming pool heating.

Source: US Department of Energy

Figure 3.10 An unglazed collector on a hotel in Mexico.

Glazed

This takes the same layout principle and adds a box, low-E coated glass and insulation.

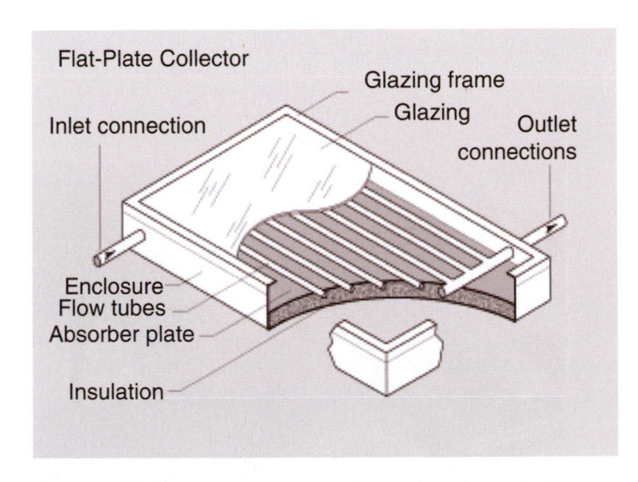

Figure 3.11 A normal flat-plate collector has the same layout but is glazed and insulated. Fins can be attached to the pipes and painted matte black to absorb the heat.

Source: US Department of Energy

Figure 3.12 Glazed flat-plate collectors in a passive system on the roof of a school in Bangalore, India.

Source: http://servamaticsolarparts.com

Batch solar water heater

Also known as the breadbox, or integral collector storage system, batch heaters include one or more tanks or tubes in an insulated glazed box. The tanks are black or painted with TISS paint. The simplest possible system involves painting a tank black, putting it in a crate, insulating it on all sides except the one pointing at the sun and covering the sun side with glass or plastic. Cold water enters the tank at the bottom. The heated water inside the tank rises to the top of the collector and is drawn into the building as needed or into a storage tank. A curved mirror can be arranged around the tank to reflect more heat onto it, and in this case the tank is only insulated at the back.

Batch heaters should be used in places where freezing is infrequent. They heat the whole tank in the collector at once so take longer to heat a little water for morning use. Since the tank is outside, it will cool down more quickly. Therefore, batch heaters are most appropriate where water is going to be used later in the day. Modern batch solar collectors come in a dome covered not with glass but a material especially designed for transmitting solar energy while holding in heat, for example twin-walled GE Lexan®. The dome and insulated base together provide protection against windchill.

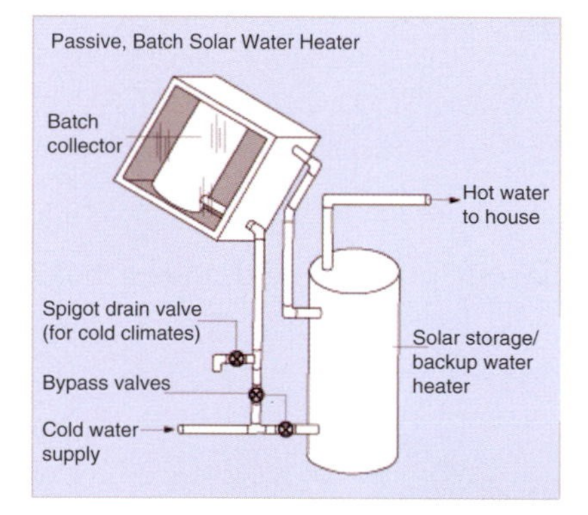

Figure 3.13 A passive, batch solar water heater.

Source: US Department of Energy

Figure 3.14 Batch solar heaters in California.

Source: © http://servamaticsolarparts.com

Evacuated-tube collectors

Evacuated tubes look more contemporary than flat-plate collectors. They consist of rows of glass tubes. Inside each tube is a vacuum (hence 'evacuated' – it's the air that's been removed) and a long collector painted with the black selective surface

Figure 3.15 Evacuated tubes.

Source: © Bosch

Figure 3.16 Tube collectors on a 211 apartment block in Wezembeek, Belgium, yielding 6.6 MWh per year.

Source: IEA-SHC

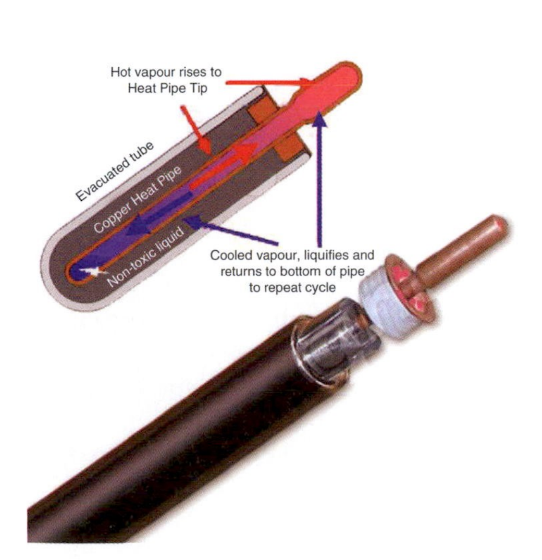

Figure 3.17 Cross-section through an evacuated-tube collector.

Source: US Department of Energy

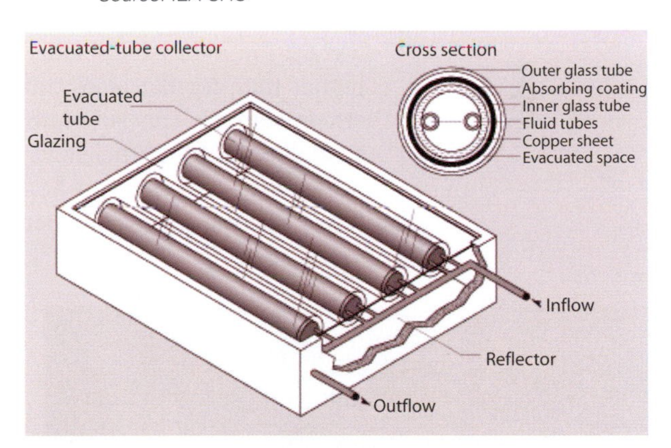

Figure 3.18 Evacuated tubes contain vacuums to retain more heat. Fins, or a cylindrical surface, that maximize the area of collector directly facing the sun are attached to, or enclose, the fluid-containing tubes. This increases the period over which heat can be collected, both during a day and throughout the year. The surface is given a coating, which absorbs more heat radiation frequencies. Evacuated tubes are more efficient than flat-plate collectors mainly because they do not re-emit as much heat thanks to the vacuum within the tubes.

Source: US Department of Energy

that prevents the infrared re-radiating out. The surfaces are tilted at the optimum angle towards the sun. The heat is conveyed to a manifold at the top of the tube where it is circulated to a heat exchanger coil in the storage tank.

There are two main types of evacuated-tube collector:

- direct flow – the fluid in the closed loop is circulated through the piping of the absorber, transferring its heat through a normal heat exchanger;
- heat pipes – the absorbed heat is transferred in a closed loop by the evaporation of the circulating fluid at the sun-heated end and its condensation at the manifold. The use of these phase changes means much more heat can be transferred.

The most common evacuated tube on the market at the moment is the Sydney tube collector, which is a direct flow collector consisting of two glass tubes fused together. The vacuum is located between the two tubes, and the outside of the inner tube is coated with the selective absorber. Inside the inner tube, the heat is removed by a copper pipe embedded in a cylindrical heat transfer fin, which conducts the heat to the manifold. There is no wet connection between the outer tube and the heat conductor, so if the external part of the tube is damaged it can be easily replaced.

Evacuated tubes or flat-plates?

In Europe, evacuated tubes occupy about 30 per cent of the market. Elsewhere in the world, the figure is more like 8 per cent, except for China. China is the market leader in vacuum tube collector production, and most tubes in the world are installed there. Because of Chinese companies' strong marketing strategies, tubes are on the increase elsewhere, too. Although the COPs for evacuated tubes are higher than for flat-plate collectors, in practice the difference is not huge in most circumstances. This is because the efficiency of the collector by itself is often only a small part of the efficiency of the entire system. What is important to the end user is the number of kilowatt-hours per year that a given system will deliver. Evacuated tubes perform marginally better, all other things being equal, in conditions where:

- roof space is limited;
- it isn't possible to use an exact south-facing orientation;
- the location is at a higher latitude;
- there is less direct sunlight – the prevailing conditions are mostly diffuse or cloudy;
- a higher output is required for space heating, too;
- there is a demand for higher temperatures.

Figure 3.19 The manufacture of evacuated tubes.

Source: © Bosch

A disadvantage of evacuated tubes is that they do not scale up as readily as flat-plates. This is because they have larger diameter manifolds, which are less conducive to fluid flow than the smaller ones in the serpentine heat elements of most modern flat-plate collectors. Flat-plate

Figure 3.20 Solar collectors integrated into the side of a social housing block in Paris, France, designed by Philippon-Kalt Architects. Seventeen separate installations supply the 17 apartments with 40% of their domestic hot water needs. The double-skinned facade also offers privacy and restricts noise.

Source: © IEA-SHC

collectors are generally preferred where they are cheaper and where there are clear skies for much of the year. Evacuated tubes perform over more of the year – including colder months. With both types of plates, bear in mind that they are potentially vulnerable to damage from hailstones, although flat-plate collectors are tested to be hailstone proof under the standard EN12975.

Evacuated tubes can work in a system with combi boilers but will require the installation of a storage tank. There are then two options: either the tank output feeds into the combi boiler, having preheated it, or bypasses it when a sensor detects it is hot enough. The option chosen depends on the exact boiler model.

Polymer-based collectors

A new generation of solar collectors is on the cusp of mass-market introduction. These are made not of glass and metal, but of one moulded piece of polymer-based plastic. The material offers a number of advantages but has also presented problems, which are now being largely overcome following research into the right kind of material. The advantages and disadvantages are as follows:

Advantages:

- up to 50 per cent cheaper;
- quicker to install;
- greater potential choice of colours and designs;
- open, direct systems are possible in warmer climates, in open systems;
- thermosiphon systems are possible in warmer climates;
- they are not vulnerable to freezing so water rather than antifreeze can be used in closed loop systems;
- smaller bore (pipe diameter) and new geometrical shapes of the absorbing layer yield a faster response and result in slimmer collectors, meaning potentially greater efficiency than flat-plate collectors;
- some models include a PV cell to power a pump where one is used.

Figure 3.21 A polymer-based panel with integral PV cell for powering the pump.

Source: © SolarTwin

Disadvantages:

- depending on the material and conditions, there is a risk of damage over the long term due to freezing, exposure to ultraviolet light, or water boiling in the panels when not in use.

The disadvantages are being overcome, in many cases, by research and development on suitable polymers, such as polypropylene or cross-linked polyethylene (PEX), enabling the products to be installed at higher latitudes. Polyphenylene sulphide blends have been found that are able to withstand sustained temperatures in the range of 170–180°C (340–355°F). Almost available on the market is a freeze-resistant collector that can work as far north as Norway.

Unglazed pool absorbers in polymeric materials have been successful for over 20 years. Materials include polypropylene, ethylene propylene diene monomer (EPDM) rubber, polyvinyl chloride/polyurethane (PVC/PU) covered polyester texture, polyethylene pipes and EPDM pipes, coloured red to match the colour of the roof tiles. One domestic heating product, which is glazed with polycarbonate, has been on the market for 11 years with considerable success. Most such available models have been designed for low-pressure systems. One absorber is made of silicone rubber tubing partly compressed between metal plates. Another is a drain-back collector with a stiff, extruded absorber plate of polyphenylene ether/polystyrene (PPE/PS) blend.

Collector covers available include twin-wall sheets of polycarbonate, dome structures of polycarbonate, and acrylic glazing. Polymer absorbers are on the market with integrated storage – batch collectors and similar. These collectors are coated with TISS, to maximize absorption of the thermal radiation, and research has established that these paints have great longevity, and can be safely used in solar collectors for at least 45 years. They have been developed to be water and oil repellent, and so need far less cleaning than glass and other collector coverings to retain the same level of performance.

Figure 3.22 This house in Aarhus, Denmark, utilizes passive and active solar heating together with photovoltaic panels. The 6.7m² (72ft²) of solar collectors cover 50–60% of hot water requirements throughout the year, as well as supplementing space heating. A heat pump satisfies the remaining 50% of the heating that is not provided for by the passive solar design. Underfloor heating is used in all rooms except for the bedrooms.

Source: © IEA

This type of panel has the potential to change the game for SWH by making it much cheaper and easier to install, and much more adaptable to different architectural styles and tastes. Research and development on polymers is happening in tandem with research on polymer-based storage tanks and heat exchangers, and polymer-based PV cells. There are large heat storage systems using high-density polyethylene (HDPE) liners for district and inter-seasonal storage, for example, for a gravel-pit water heat store.

Solar water heating systems

There are several different systems with which these collectors can be integrated. SWH systems can be either active (pumped) or passive (thermosiphoned), indirect (loop) or direct (water heated is water used). Thermosiphoning involves the circulation of the water by convection.

Table 3.1 Open and closed, active and passive solar water heating systems.

	Direct	Indirect
Active	pumped, water heated is water used	pumped, heat is transferred from a closed loop
Passive	thermosiphoned, water heated is water used	thermosiphoned, heat is transferred from a closed loop

Passive (commonly referred to as thermosiphon)

Advantages

- does not require a pump;
- more reliable as there is less to break down.

Disadvantages

- the storage tank must be situated above the top of the collector(s) or thermosiphoning will not work;
- the thermosiphon effect only works consistently well in warm climates.

Active

Advantages

- the tank can be located anywhere.

Disadvantages

- requires a pump and electricity source;
- requires a valve or thermostat controller so that the pump only comes on when the collector is hotter than the tank.

Direct

Advantages

- easiest to install;
- used in tropical settings where it never freezes;
- also useful for swimming pools;
- low maintenance.

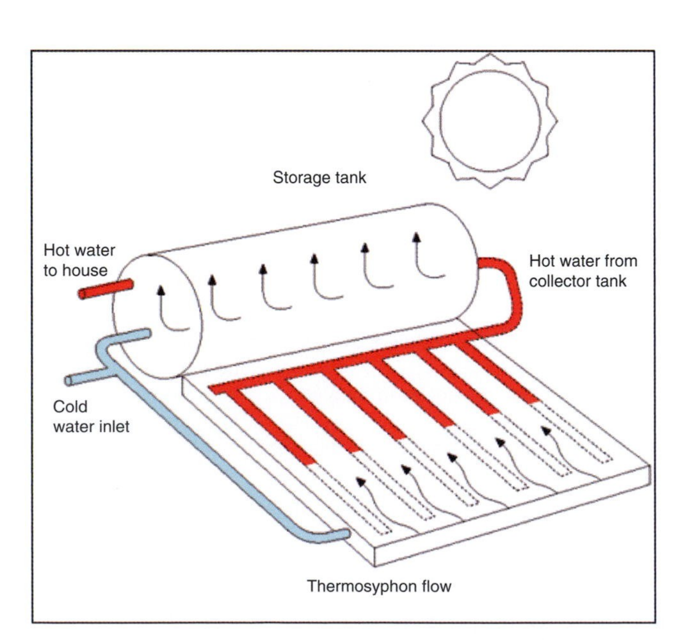

Figure 3.23 Schematic diagram of a direct, passive system.

Source: US Department of Energy

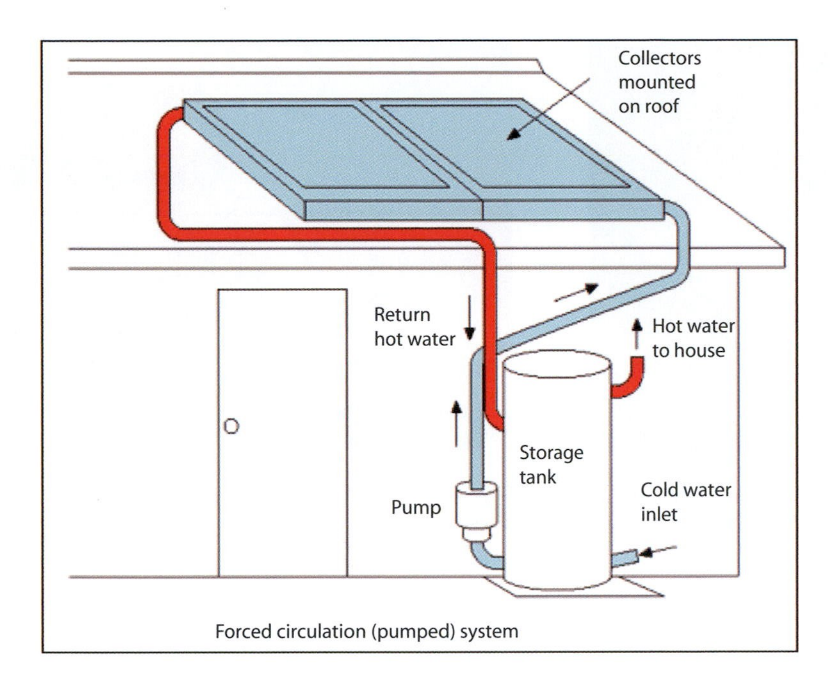

Figure 3.24 Schematic diagram of an active direct system.

Source: Government of Western Australia

Disadvantages

- unsuitable in areas where it freezes unless the system is drained;
- if the tank is outside then thermal losses will be high;
- the water heated is the water used.

Indirect

Advantages

- can be used in areas where the temperature drops below freezing;
- the water being heated indirectly can be in an insulated tank indoors so thermal losses will be lower than for the direct system.

Disadvantages

- more complex and requires a heat exchanger;
- this negatively affects the efficiency of the system;
- the heat-transmitting fluid (which contains anti-freeze) must either be non-toxic or there must be no chance of it leaking into the water.

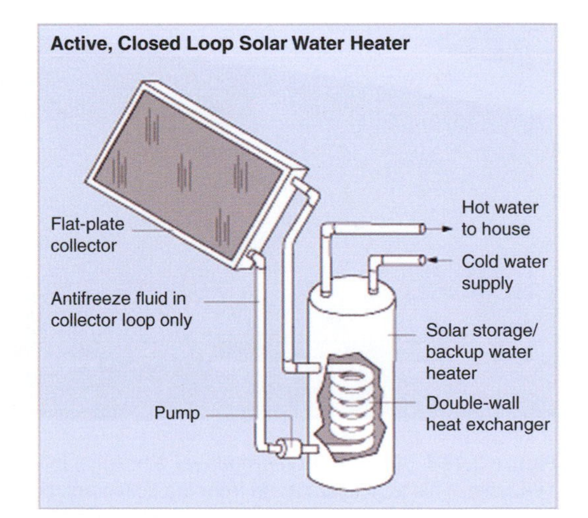

Figure 3.25 Schematic diagram of an active, indirect system.

Source: US Department of Energy

Thermosiphon system

The simplest system, which is found in most countries with a latitude of within 20° from the equator, is to have

Figure 3.26 Thermosiphon solar water heating kit with integral tank used in the Cirque de Mafate, Réunion.

Source: Wikimedia Commons

Figure 3.27 Thermosiphoning batch system on the roof of a house in Spain.

Source: Wikimedia Commons

the storage tank directly above the collector on the roof of the house. Water circulates by natural convection because hot water rises and cold water falls. This method is called thermosiphoning. It does not need a pump. The system is known as passive (or 'compact') for this reason. The tank can be an integral unit with the collectors or it can be located away from them, inside a roof space to help the water stay hot for longer. It doesn't matter as long as it is positioned higher than, and close to, the top of the collectors, allowing the heated water to rise by natural convection into the tank.

This type of system is robust and long-lasting. There is little to go wrong. Systems that were installed in Colombia in the 1970s and 1980s are still working. Here, a man called Paolo Lugari developed designs under the company name Las Gaviotas, adapting the best systems from Israel. Colombia's Banco Central Hipotecario (BCH) stipulated that the system must be operational in cities like Bogotá, where more than 200 days in the year are overcast. The systems came with a 25-year warranty and over 40,000 were installed. For places like Bogotá, where not every day is sunny, solar water heating cannot provide all of the hot water needed throughout the year. In these cases, the solar heated water is stored in a tank, which is connected to another heating source, such as an electric immersion element or a wood burning stove, for example. The sun heats the water up to a certain temperature, and so the auxiliary heating source has less work to do to bring it up to the desired temperature.

Figure 3.28 A thermosiphon system for a hospital in Wasso, Tanzania. This tank is separate from the collectors, but notice how it is completely above them to enable the thermosiphon effect. It is tall and narrow in order to promote temperature stratification: the hottest water naturally convects to the top ready to be drawn off.

Source: © Frank Jackson

Direct system design

A direct system is most often installed in sunbelt climates, such as the central and southern areas of Florida, and the water heated is the water used. A standard tank can be used, connected to a 3.7m^2 (40ft^2) solar thermal collector. The tank provides storage for preheated water that feeds an auxiliary water heater. If the tank has an electric immersion coil, normally the element is not connected. An air vent is installed at the high point of the solar thermal collector to purge air when the system is initially filled.

The pump – a small circulator pump using as little as 10W – can be powered directly by a 10W PV module, or a thermostatically controlled AC pump can be used. A thermostat can be installed to limit the temperature the solar tank reaches for safety purposes. The cold water feed enters the tank at the very bottom. From here it is drawn to the collector to be heated. It then returns to the tank one-third of the way down from the top. The water to be used is drawn from the very top of the tank; thus the water in the tank is heat-stratified. Heat is not being wasted by already heated water being circulated back to the collector, and only the hottest water is being used, which naturally rises to the top of the tank.

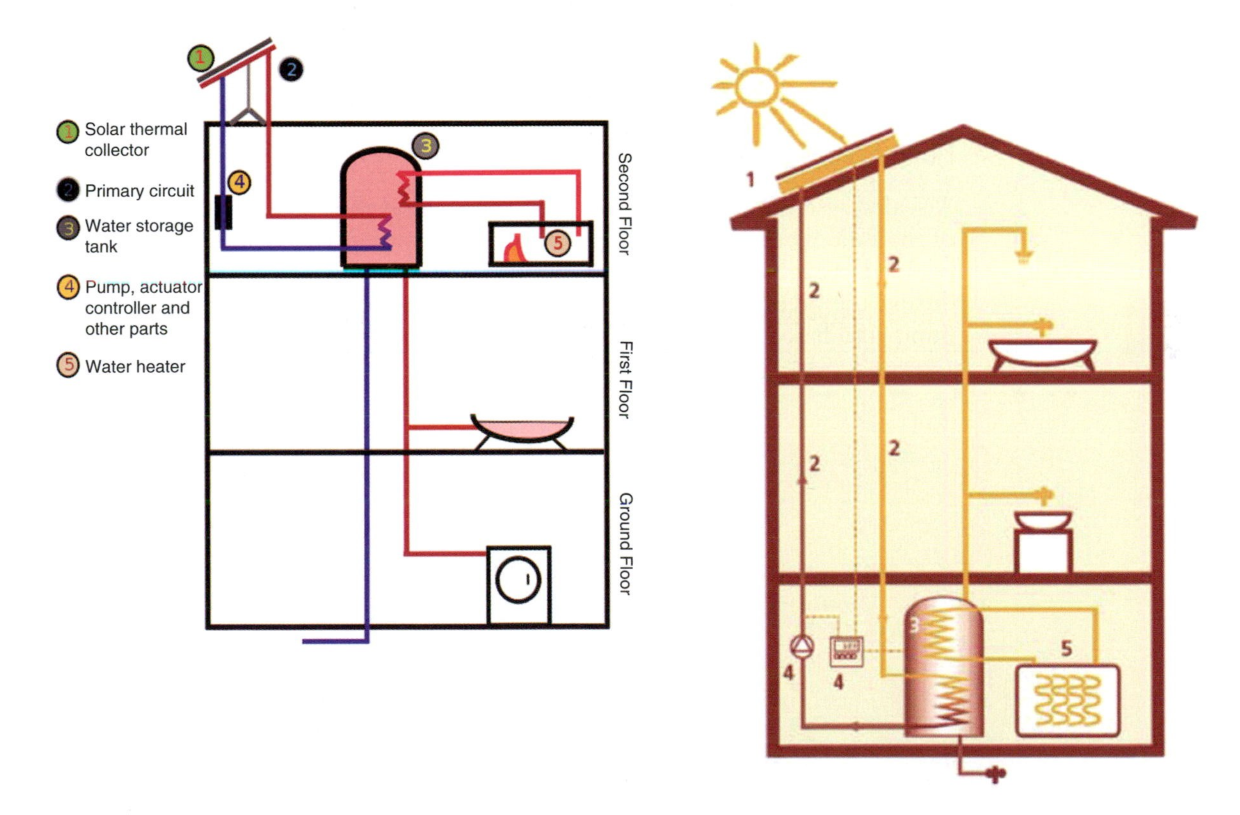

Figure 3.29 Schematic diagrams for an indirect, active solar water heating system. As the tank is below the collectors, a pump must be used. The pump in the building on the left must be more powerful as it has further to pump the water, therefore it is better to position the tank closer to the collectors, as on the right, and let gravity feed the water to the point of use. It is called 'closed loop' because the liquid is returned to be heated again, after transferring its heat to the water in the tank.

Source: Energy Saving Trust

Indirect, active solar water heating system

The closed loop active system is the most commonly used in cooler countries. It consists of the following elements:

1 The solar collector absorbs the incident solar energy. The heat is conducted into a heat-carrying fluid, usually water with antifreeze – propylene glycol (this is non-toxic, unlike the ethylene glycol that is used in vehicles) – that passes through the collector.
2 The primary loop circulates the heat down to the heat exchanger coil in the storage tank and back up. The water in the tank absorbs the heat from the heat exchanger coil and the cooled fluid returns to the collector to start the process again.
3 The heated water is stored in an insulated tank.
4 A controller controls the pump to only come on when the collector is hotter than the tank. This is to prevent the pump from circulating the heat from the tank back up to the collector at night. A one-way valve also prevents it thermosiphoning back up with the same result.
5 Auxiliary heating, such as a gas or biomass boiler or electric coil, can be used to top up the heat in the tank if required.

In some countries, there is a choice between two common primary system layouts: fully-filled and drainback.

Fully-filled systems

A sealed system must include an expansion vessel that takes up the expansion of the fluid when it gets hot. The pipework may undulate as long as air can be released at high points and fluid drained at low points. Undesirable heat loss during the night is prevented by a one-way check valve positioned after the pump and before the expansion vessel and the solar collector.

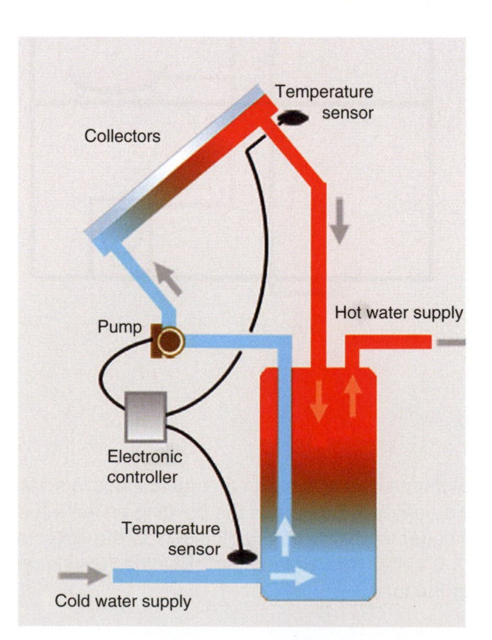

Figure 3.30 Direct system design for areas where freezing is rare. The controller compares readings from the two temperature sensors and switches on the pump when the collector water is hotter than the return from the tank. The water heated is the water used.

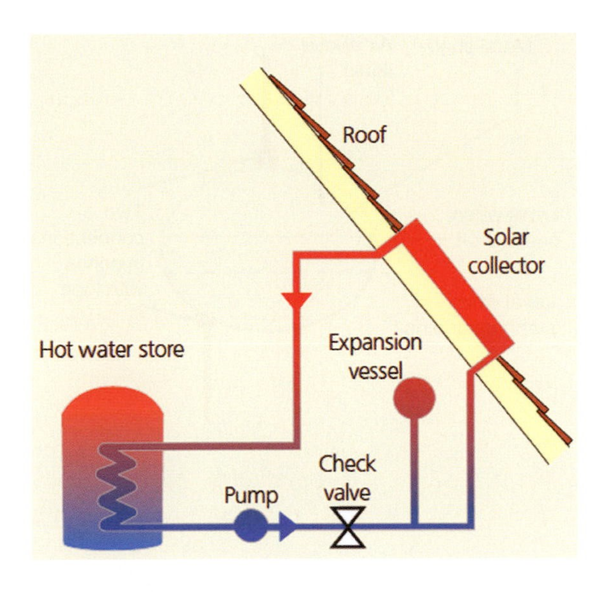

Figure 3.31a An indirect, sealed solar thermal system layout. It includes an expansion vessel to contain any expansion of the fluid when it gets hot.

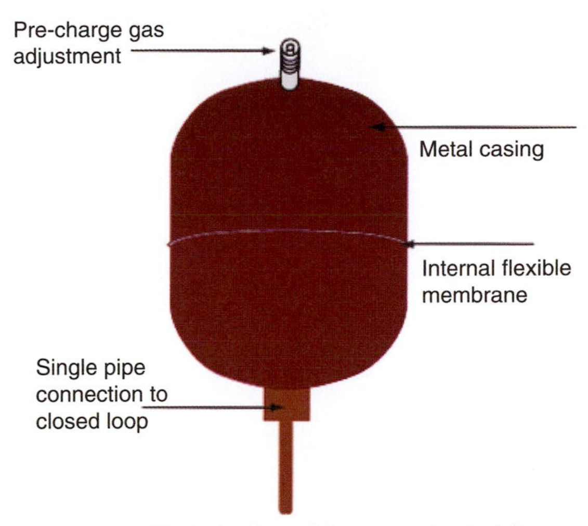

Figure 3.31b The behaviour of the expansion tank in an indirect, sealed solar thermal system. The vessel is not insulated because, if hot fluid expands into it, it is desirable for it to cool down as quickly as possible.

Drainback systems

With a drainback system, air is present in the circuit and any expansion of the fluid is contained by the drainback vessel with an air pocket. Undesirable heat

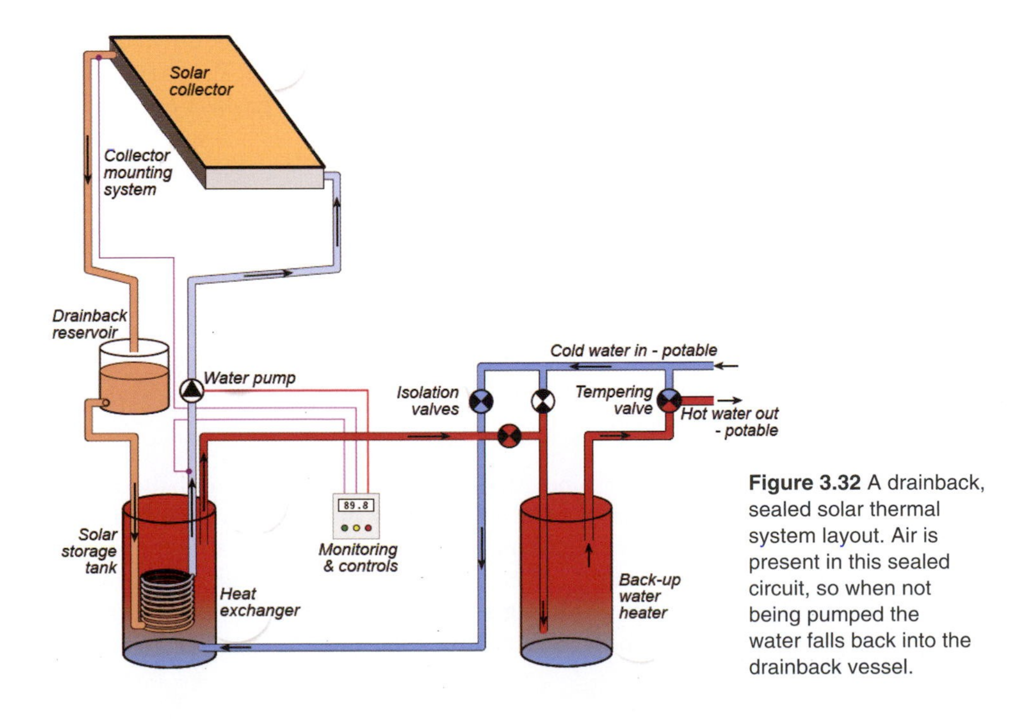

Figure 3.32 A drainback, sealed solar thermal system layout. Air is present in this sealed circuit, so when not being pumped the water falls back into the drainback vessel.

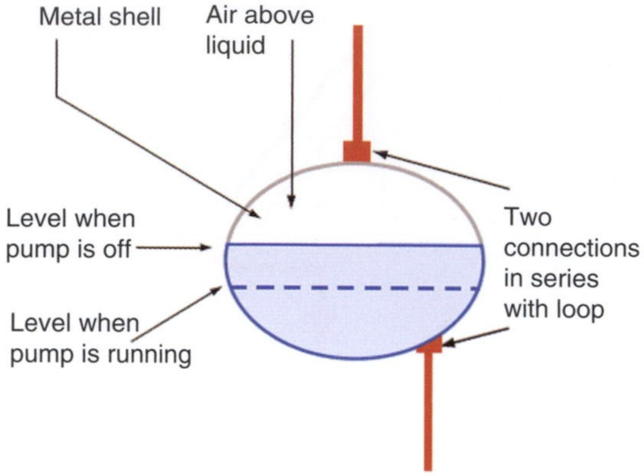

Figure 3.33 The drainback vessel in a drainback solar thermal system. It is insulated as it is part of the loop and gets hot.

loss is prevented by switching the pump off, which means that the fluid drains back from the collector into the drainback vessel positioned after the collector in the circuit. It requires a more powerful pump than a fully-filled system.

A drainback system has two main advantages over a fully-filled system:

1 There is no risk of overheating because when the collector gets too hot the pump switches off and the fluid drains back out of the collectors and into the drainback vessel.
2 There is no risk of freezing because, again, the pump is off when the collector gets too cold.

Repeated overheating of the glycol fluid causes it to break down chemically and produce a sludge which can quickly block the system. Propane fluid might be used instead of glycol, which might corrode plastic gaskets over time.

Choosing the appropriate system

For small or domestic systems the basic questions to ask are:

- how much unshaded equator-facing roof space is available? (a northerly aspect in the southern hemisphere, southerly in the northern hemisphere);
- what is the nature of the existing water heating system? (for example, some combi boilers may not be suitable);
- what is the budget?

All existing hot water pipes should be well insulated and the most efficient showerheads taps, washing machines and dishwashers installed, in order to minimize the hot water requirement in the first place. It is far more cost efficient to minimize the demand for energy than it is to invest in renewable energy technologies. When considering solar space heating too, make sure to read the section later in the chapter on passive solar design.

Collectors can be up to 30° east or west of an equator-facing direction (tubes are more tolerant of this than flat-plates), and within the site's latitude plus or minus 15°; for example if the latitude is 20° then the allowable range is 5–35°. Ideally the spot should have no shading between 9am and 3pm. If a roof is not exactly equator-facing, the area of the solar collectors may need to be increased to compensate. This might also be the case if there is unavoidable shading going across say 10 per cent of the collector area at certain times during the day or the year. Collectors must be positioned as near as possible to the tank to minimize heat losses. The pipes must be very well insulated with no gaps or breaks in the insulation.

Figure 3.34 Flat-plate collectors on a house in the UK.

Source: © Energy Saving Trust.

System sizing

The further the site is from the equator, or the cloudier the climate, then the greater the surface area of collectors that will be required to obtain the same amount of heat. Within the tropics, for example, one square metre per person is sufficient, whereas in latitudes 50–60°, the figure is 2.5–3m² per person. A four-person household, for example, uses slightly less than 5000kWh per year for hot water. In a climate such as Canada's, a normal solar collector of the flat-plate variety could produce around 400kWh per square metre per year in a typical well-maintained system. As the domestic hot water demand stays more or less the same throughout the year, this is an easy way to calculate the size. An estimate of the available energy at a site and therefore the size of collector area and storage volume can be derived from inputting the latitude and longitude, angle of incidence of the slope of the roof and the orientation of the roof towards the sun, into freely available software, which is described in the Resources chapter. As a rough guide to the amount of collector area required (without taking into account the angle of incidence or amount of cloud cover in the vicinity), Table 3.2 may help.

Table 3.2 indicates that a household of four in Arizona (30° latitude) would need about 4m² (44ft²) of collector, but the same household in Sweden (55°) would need around 11m² (120ft²).

Storing the heat

Usually, 300 litres (80 gallons) of hot water storage is sufficient for four people. The volume of the storage tank can also be calculated from the solar collector area. Typically, for every square metre of solar panel about 50–100l (15–25 gallons) of storage volume is appropriate, depending on the solar resource at the location. So, for example, for a system with 5m² of solar panel you might need a tank capable of storing around 300–375l.

Table 3.2 Collector area, assuming a requirement of 75 litres of hot water per person per day.

latitude	collector area per person (m²/ft²)
0°	0.5/5.5
20°	0.75/8
30°	1/11
35°	1.3/14
40°	2/22
50°	2.5/27.5
60°	3/33

Once captured, the heat wants to leave, since, according to the second law of thermodynamics, heat will always try to spread to a colder place. The speed at which it does so depends on the temperature difference between the hot water at the top of the tank and what is adjacent to it: the inside surface of the tank and the cooler water below. It also depends upon the degree of convection occurring within the tank. We can limit these variables. The first is simple: by wrapping as much insulation around the tank as possible. The second is, in practice, dependent on the available space, the budget and the climate, as discussed below. The third variable – convection – is affected by the positions of the coils, any electrical heating element present, the location of the outlet and inlet pipes, and the temperature of the water at these points. Systems to control this are called stratification devices.

Hot climates

If only one tank were used, under sunny conditions it would rapidly reach a maximum temperature throughout, close to that of the collector fluid temperature. After this it could not get much hotter. If the collector fluid temperature were 82°C (180°F) and the absorber plate temperature 93°C (200°F) the rate of heat transfer would be negligible. This means we would be losing the possibility of capturing more heat than the approximately 4kWh this may contain. Might we compensate by having a larger tank? If so, the tank might not be fully heated at other times. In this case, stratification occurs: the water in the tank will contain levels, with the warmest at the top and the coldest at the bottom. Heat will always want to leak from the top level to the bottom. Therefore, if the tank is too big for the size of the collector then the collector will not operate at maximum efficiency. If the tank is too small, some of the heat collected will be wasted. The storage volume needs to be matched to the collector area for as much of the year as possible with minimal heat wastage. In both cases, to collect more heat we would need more tanks.

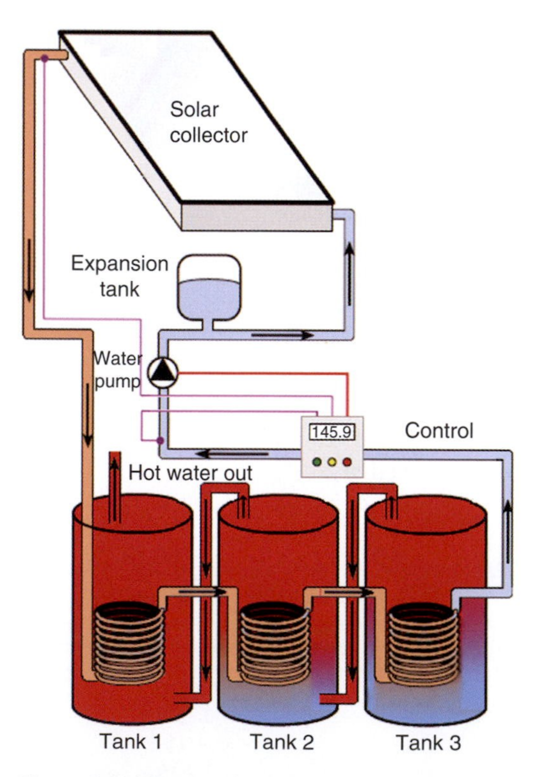

Figure 3.35 The use of two, or, in this illustration, three, well-insulated tanks in sequence, maximizes the output of the collector.

Suppose we had two, or even three, well-insulated tanks, in a sequence. The closed loop from the collector passes through these tanks. As it does so, it transfers its heat successively to the water in each tank. The first tank will start off being the warmest, since it is first in line to receive the heat. Tank two will receive the heat in the loop left over from tank one. Tank three will collect the last bit of heat. The heat transfer fluid is then recirculated back into the collector to gather more heat and start the process again. This way, the water in tank one reaches the maximum temperature. Any further heat from the collector is then absorbed by the second tank. And so on. But crucially, the heat in the first tank will not be lost to adjacent water that is cool, as would be the case if it were in a larger, single tank. The preheated water in tank three is fed into tank two, and the preheated water into two is fed into tank one, and the water used is taken from the top of tank one.

Temperate and cool climates

When *sealed, indirect* systems are installed in temperate and cooler climates, unless the property is being designed or renovated from scratch to accommodate the three-tank idea, lack of space usually results in the choice of a twin-coil cylinder or tank. Various designs of tank are on the market, with varying degrees of sophistication to maintain stratification and maximize efficiency. Near the bottom of the tank in Figure 3.36 is the heat exchanger coil in the circuit that returns to the collector. This pre-heats the water, causing it to rise. In the top half is another coil. The fuel source for this coil may be electricity, gas or a biomass boiler. This source is controlled by a thermostat, and it is

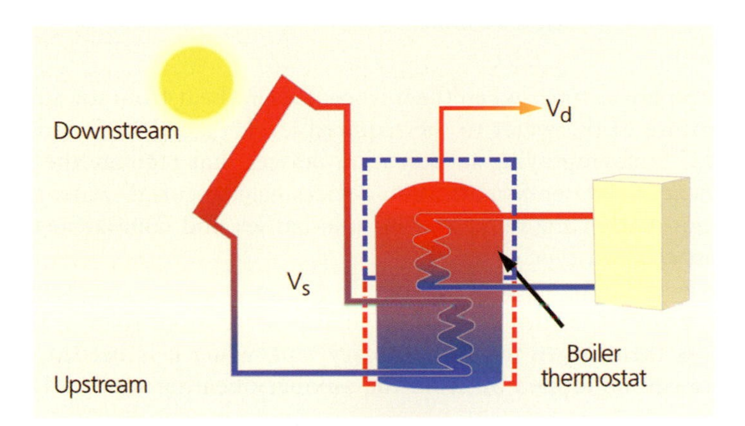

Figure 3.36 A basic twin-coil sealed, indirect single tank system.

Source: © Wagner Solar UK Ltd.

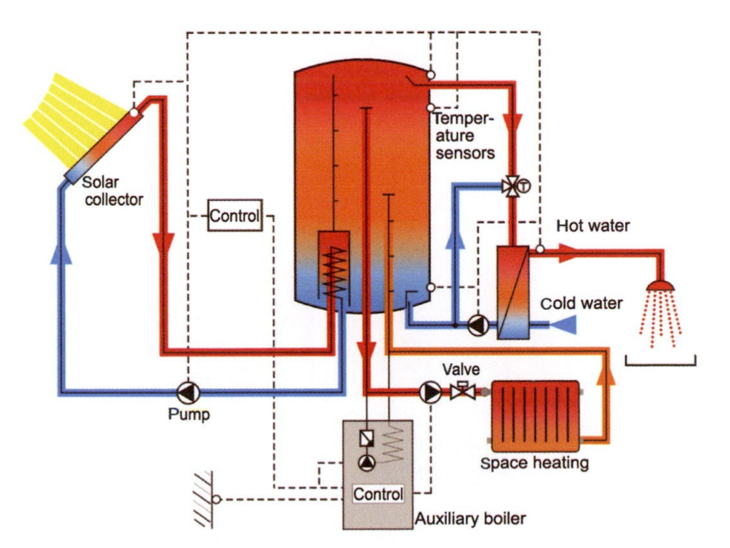

Figure 3.37 Another sealed, indirect single tank system utilizing a twin coil, but with stratification devices. Notice how the hot water is taken off at a higher point in the stratified water levels, the space heating is taken from a lower level, and returned to a lower level still. Sensors detect the temperature at different outputs and different levels within the tank, and apply the heating, remove the water, or return the water, at different levels accordingly, to maintain the stratification and maximize efficiency.

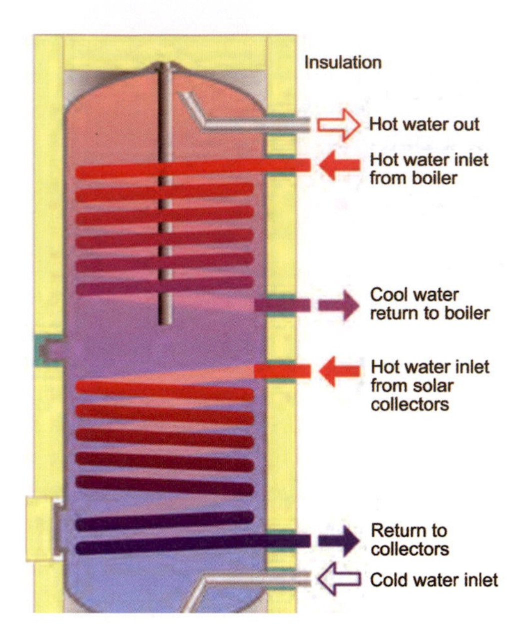

Figure 3.38 A simpler layout for stratification, without sensors.

brought into play at times when there is insufficient heat from the sun to bring the temperature of the water to the required level. Figure 3.37 shows a more sophisticated tank employing stratification devices that regulate the input and output of heat and water. Stratification devices help to curtail water movement by convection within the tank, and include baffles and compartments. These are the most efficient type of tank.

If there is sufficient space in the building, two or more tanks (Figure 3.39) can be employed. The first is a buffer tank to store more solar-heated water. This water is then drawn into the primary tank when it is needed, where its temperature can be topped up from the auxiliary heating source. This means

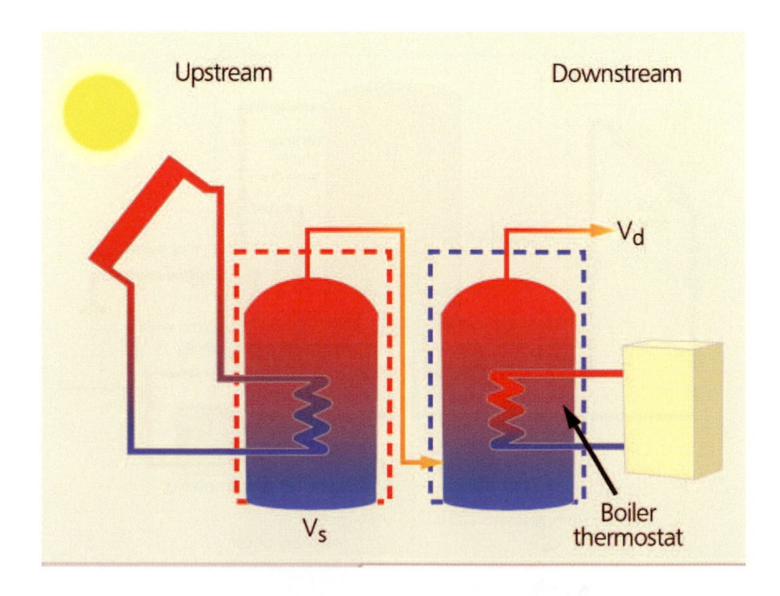

Figure 3.39 A twin tank, sealed, indirect system. Solar heated water is stored in a buffer tank, and fed into a second tank, which can top up the temperature using a secondary heat source if required.

Source: © Wagner Solar UK Ltd.

Figure 3.40 These two tanks are part of a system designed to yield hot water and space heating. The tank on the left holds hot water for consumption. The tank on the right holds further water at a lower temperature for space heating. The controls prioritize the hot water for consumption. A secondary heat source can top up either or both tanks if required. This system has been monitored and as a design solution is not recommended as it is not the most efficient.

Source: © IEA

that this secondary source has to expend less energy heating up the water. There are single tanks available which combine a domestic hot water tank situated inside a heat store – in other words, a tank-within-a-tank. This can be used if space is at a premium.

If the property already utilizes a combi boiler, multipoint, single point or electric point-of-use heating appliance, then the auxiliary heater or boiler can be supplied by the water from the solar heater tank, as if it were coming from the cold mains. When domestic hot water is drawn off the system, preheated water is fed into the boiler. However, specialist advice is needed with such a system, because only some boilers are designed to take preheated water. The thermostat must be set so that the incoming water temperature is not so hot that it damages the boiler.

Possible risks

Water temperature over 80°C (176°F) can cause severe burns instantly, or death from scalds. It is easy to be scalded by solar heated water unless the system has a thermostatically controlled mixer valve to add cold water to it when required in hot weather. The outlet from the solar preheat store should not exceed 60°C (140°F) and should have an adjustable safety control device. This is also necessary to prevent scaling, which can particularly occur at these high temperatures and eventually clog up the system.

Another potential danger is the formation of bacteria in the stored solar heated water. This risk can be minimized by:

- keeping the volume of preheated water below a figure that is twice the average daily hot water use;
- using indirect primary circuits;

- thermostatically controlling the output temperature to 60°C (140°F);
- keeping to a minimum the length of uninsulated secondary pipework.

Pump and controls

Controls and pumps are not required in batch heater systems, where the water is moved by simple water pressure, or in thermosiphon systems, where it is moved naturally by heat rising. However, in all cases, an isolation valve is essential to shut off the collector from the tank for maintenance, while allowing any auxiliary water heater to continue to supply hot water.

Pumps are used in active systems. The choice of pump is important. It must be sized correctly depending on the size of the system and the distance and the height between the collector and the tank. A direct current (DC) pump could be powered from a PV collector. A pump powered by alternating current (AC) can be plugged into a wall socket. Pumps can last up to 20 years. It should be efficient so that its power requirements are no higher than 2 per cent of the peak thermal power of the collector. Usually this means 10–30W but lower consumption pumps designed specifically for solar hot water applications are now on the market.

The pump and system are governed by a differential thermostatic (or temperature) controller (DTC). This includes temperature sensors on the collector and on the tank, and compares the readings from both. It allows precise adjustment of when the pump is switched on and off, and provides a display for commissioning and fault detection. It should have a manual override. The primary system should also include a pressure gauge to show the pressure of the system and thus to alert for leaks (if the pressure drops). Faults in a solar water heating system can remain unnoticed by the user because the auxiliary heat source will automatically compensate for inefficiencies.

System installation

Hopefully, a SWH system will be installed in an energy-efficient building. Such a building – whatever the climate – will be wrapped in layers of insulation and airtightness and possibly wind-tightness barriers. As the pipes to the collectors penetrate the walls or roof, and therefore the insulation and barriers, great care must be taken to ensure that flashings, tapes and insulation is used to seal holes and prevent water and air infiltration. In some structures, there is also a risk of interstitial condensation (between building fabric layers), where warm humid air meets the cold return pipe to the collector. Water may then condense on it and drip down into timbers causing rot. Care should be taken to prevent such humid air reaching this point, without compromising the ability of the fabric to breathe where it needs to.

Mounting the collectors

Collectors can be mounted on a roof, on the ground or on an awning. Roof-mounted collectors are sometimes on a frame parallel to and slightly above the roof. Ground-mounted systems should be vandal proof and can be tilted to the optimum angle. An awning mount will attach the collectors at the correct angle to a vertical wall facing in the appropriate direction if a roof space is not available.

Figure 3.41 Practising installing solar collectors on a roof during a training course.

Source: © Chris Laughton

Roof mounts are usually the cheapest option, if the tilt and orientation are acceptable. Structural loading must be considered, and in areas exposed to high wind and snow the effect of this must be factored in. If weight is an issue, then ground mounts would have to be used. On the ground, there is a higher risk of shading.

Maintenance

Most professionally installed systems come with a 10-year warranty and require little maintenance apart perhaps from occasionally cleaning the collectors. A yearly check, and a more detailed check every three to five years by a competent person should be sufficient.

System expectations

In temperate and sub-arctic zones, the systems described above are designed so that in summer they can completely meet all the water heating requirements. In winter, most water heating will need to be supplied by another source, whether it is gas, oil or biomass, supported by solar thermal heating whenever it is sunny. On average, it can mean that, throughout the year, around 60 per cent of water heating requirements can be met by the system in medium latitudes and 40 per cent in higher latitudes (depending on collector area and system efficiency). Closer to the equator, almost all water can realistically be solar heated throughout the year.

On a typical summer day in a temperate zone, the fluid in the collector can reach up to 80°C (180°F). On cloudy and warm days, it can reach 20–30°C (70–90°F). On cloudy and cooler days, it can reach 10–15°C (50–60°F). But as long as the temperature in the collector is higher than that of the incoming cold water (usually about 10°C (50°F), then a solar water heating system will save energy.

Dual systems for water and space heating

Some companies supply dual systems, which provide space heating as well as hot water. The solar collectors have to cover a much larger area. They supplement the use of another fuel source to heat rooms during the autumn and spring. Depending on how well the building is insulated, and its location, a solar thermal system can supply 20–30 per cent of the heating demand. If combined with passive solar building design, it is possible to supply all of the heating requirements from the sun.

See the section on active space heating for more information.

(a)

(b)

Figure 3.42 Flat-plate and evacuated-tube collectors on rooftops in Berlin; the tubes are on the roof of a government building.

Source: © (a) BSW-Solar/Langrock. (b) BSW-Solar/Upmann

Figure 3.43 A test rig trickle-down collector with a pump powered by a PV module.

Source: © John Canivan

Solar electric-thermal collectors

There are new, hybrid solar collectors on the market that incorporate both PV power generation and solar water heating (thermal) – called PV/T. These are discussed on page 140.

Trickle-down solar collectors

Trickle-down solar collectors are a different concept from SWH collectors. Invented by Harry E. Thompson in the 1950s, in their basic conception they are simply a black, corrugated metal (aluminium) roof covered with glass. The back is insulated. Water trickles down the troughs in the corrugation and is heated up on the way. It is collected at the bottom and then passed into a tank for storage until it can be used – either for space heating or consumption. The water used is that which is heated directly. The efficiency is therefore around 60 per cent. It is a do-it-yourself solution, but it is simple and cheap.

This kind of system is ideal in places where there is a great deal of sunshine. Since its invention, there have been refinements, which have made it applicable in higher latitudes. John Canivan uses, instead of corrugated metal, a black polyester material which distributes a thin sheet of water that therefore absorbs more heat. Instead of using fragile and heavy glass, the glazing is a polycarbonate material with an inner polypropylene film to prevent the heat escaping, although low-E glass could also be used.

In both systems, a pump is used on demand to circulate the water. When the pump shuts off, water drains back into the heat storage tanks. Kits can be purchased online (see *MTD Solar Heating*, self-published by John J. Canivan, 2007, US).

Advantages

- no problems with potential high pressures in the system causing damage;
- no airlocks can occur;
- almost half of the black surface area is directly in contact with the water, increasing heat transfer; some modern plastic encapsulated designs of solar collectors also have this feature;
- construction requires only moderate skill levels;
- commissioning and testing it is relatively straightforward.

Swimming pool heating

For those lucky enough to have an outdoor swimming pool, and for outdoor public or leisure centre lidos, solar heating is an excellent way to cost-effectively extend the period over which it can be used. Typically, the water loop through the collector is simply connected to the pool's filtration system, directly heating the water.

How much collector area is required? As a rough guide, in temperate countries, an area about half the size of the pool needs to be available on an equator-facing aspect near to the pool. In Canadian or Scandinavian climates (where they are frequently used), about an equivalent area is required. In a Californian climate, one-quarter to one-third of the area would be sufficient.

Figure 3.44 A swimming pool heated from rooftop panels that occupy about half the surface area that the pool occupies.

Source: © UMA Solar

Various types of collector are available: one popular solution is the PDM-quality PVC matting, but there are severe environmental doubts about this material. Several companies supply easy to assemble kits. Options for small pools include a black plastic matting that can be rolled up when not required; at the edge of the matting, pipes connect to the filtration pump. Ordinary solar panels may also be used. Most solutions are modular, allowing systems to be expanded with more collectors if insufficient heat is generated for a given time of year.

The amount of collector area required will be reduced if the pool is well insulated and the surface is covered with an insulating layer when not in use, to keep in the heat. A cover will also keep leaves and dust out of the pool. If the cover is black, it will also absorb more heat in the daytime. Domestic waste piping must not be used for this or any other pumped system, as it is not designed to accept the water pressure produced by the pump. Domestic pool solar collectors are usually designed to receive up to 1.5 bar; pool filtration is typically +/–1 bar. Stainless steel pipe is also not appropriate for use with swimming pool systems because the chlorine in the water causes it to corrode.

Solar cooling

Counter-intuitive as it may seem, it is also possible to cool water and buildings using solar thermal power. Millions of air conditioning units are in use around the world, the growth markets being the US and Asia in particular. Global demand for Heating, Ventilation and Air Conditioning (HVAC) equipment is projected to rise over 6 per cent per year through 2014 to a market size of more than $88 billion. Even in Canada, which has a much cooler climate, 38 per cent of housing units have air conditioning, mostly in Ontario. The electricity demand associated with air conditioning and the chilling of water, particularly during heat waves, is massive, and contributes considerably to peak loads; it has been implicated many times in blackouts as local grids are overloaded. From the point of view of solar technology, this is a hugely untapped potential market that could contribute greatly to reducing reliance on fossil fuels. Moreover, there is a great match between the energy supply and demand: it is when the sun is shining the most that there is the most power available, and the demand for air conditioning is at its peak.

Once more, Augustin Mouchot is a pioneer in this field having, in 1878, produced ice from his solar dish using a periodical absorption machine developed by Edmond Carré. The principle of absorption cooling is still used along with other thermally-driven cooling technologies, such as adsorption cooling and solid and liquid desiccant cooling. What they all have in common is that the external energy required to drive the process can come from solar heat. The absorption system works as follows: a heat-driven concentration difference moves the refrigerant vapours (usually water) from the evaporator to the condenser. At the evaporator end of the cycle, the refrigerant is highly concentrated. It absorbs refrigerant vapours (which, of course, dilutes it). It is then exposed to the solar heat, which drives off these vapours, to increase the concentration again. Lithium bromide is the most common absorbent used in commercial cooling equipment, with water used as the refrigerant. Smaller absorption chillers sometimes use water as the absorbent and ammonia as the refrigerant. Absorption must operate at very low pressures.

The market availability of absorption chillers is mainly applied in combination with district heating or heating from cogeneration. The output of solar-thermal chillers ranges from 10kW to 5MW, and can provide cold water as well as air conditioning. However, there are only a few in the range below 100kW. The COP of these models varies between 0.7 and 1.1. The 50–400kW range is generally served by adsorption chillers operating with a solid adsorbent such as silica gel. The input temperature of about 60–90°C (140–194°F) is provided by flat-plate or vacuum tube collectors with a coefficient of performance of 0.5–0.7. They cost considerably more than absorption chillers.

Absorption chillers, with a liquid absorbent, cover the 15kW to 5MW range. Input temperatures on these systems are in the range of 80–110°C (175–230°F) with a COP of 0.6–0.8. Absorption chillers need about 3–3.5m^2 of collector surface per kilowatt of cooling capacity. Desiccant and evaporative cooling systems achieve cooling capacity in the 20–350kW range. They can be expanded modularly. The operating temperature is only around 45–95°C (113–203°F); this means that the heat that can be provided by simple flat-plate collectors and in some cases even air collectors, with a COP of 0.5–1.0. As a rough guideline for open cycle desiccant systems, about 8–10m^2 of current area can be assumed necessary per 1000m^3 per hour of installed capacity. These supply cooling and dehumidification, and are widely available.

Systems with rotating sorption wheels (sorption rotors) are the most common at present. The sorption wheel contains small air channels with a large surface contact area that has been treated with a hygroscopic material such as silica gel. This dehumidifies the incoming air in the first part of the rotor. The

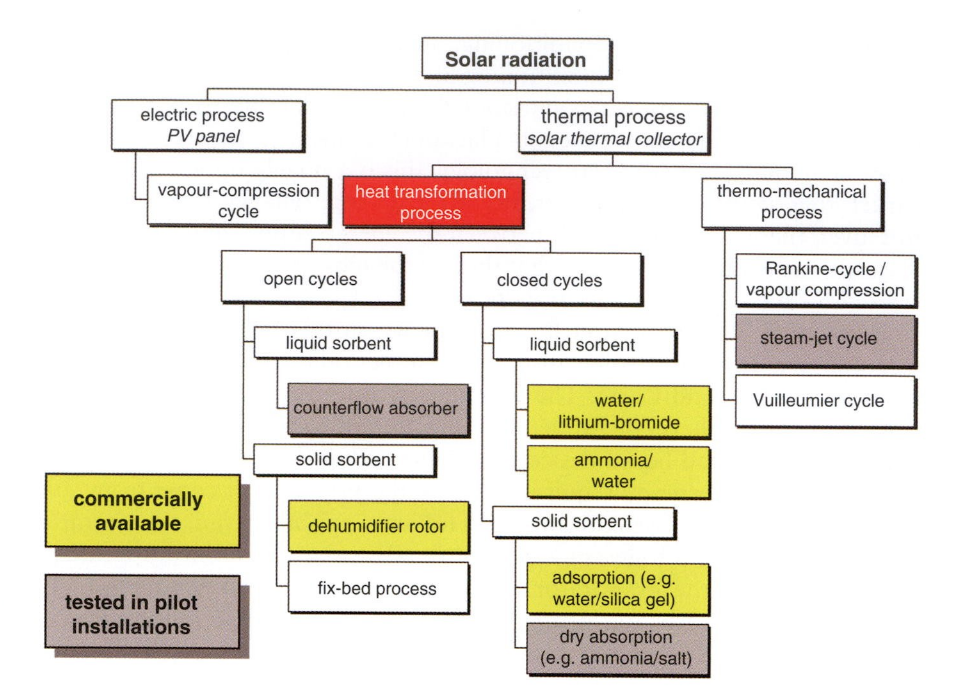

Figure 3.45 Classification of solar cooling systems.

Source: © ISE

air is then heated through adsorption by the exhaust air in the second part of the rotor. Finally, the air is cooled down in a heat reclamation rotor by conventional evaporation and humidification. There are up to ten manufacturers of sorption wheel systems around the world. In the near future, commercially available thermally driven cooling equipment in the range below 20kW is certainly possible, when combined with solar heating for space and water.

So far, there are only around 300 examples of solar cooling; the technology is still in development and standardization is in process. In most cases, solar-assisted cooling is not seen as economic without subsidy, but there is a large

Table 3.3 Solar cooling technologies and their requirements.

Cooling technology	Temperature requirements	Appropriate collectors
Desiccant systems	50–80°C (122–176°F)	solar air collectors, flat-plate collectors
Adsorption	65–85°C (149–185°F)	flat-plate collectors, evacuated tubes
Single-effect absorption (conflict air conditioning)	70–100°C (158–210°F)	flat-plate collectors, evacuated tubes, optical concentration without tracking
Double-effect absorption	130–160°C (266–320°F)	highly efficient evacuated tubes, optical concentration with tracking

Source: ISE

potential in primary energy saving. There is enormous potential for this technology.

There are several excellent example systems already installed. For example, in Dalaman, Turkey, a system which uses a 180m^2 array of tracking parabolic trough collectors has an operating temperature of 180°C (350°F) and powers a two-stage absorption chiller with a COP of 1.2–1.5. It provides air conditioning for a hotel and supplies steam for a laundry. In Italy, the European Academy (EURAC), uses 480m^2 of evacuated tube collectors to power a 300kW absorption chiller and a 630kW compression chamber. The system includes 10m^3 of heat storage and 5m^3 of cold storage and has been in operation since 2005. The largest system in the world at the time of writing, is located in Greece, at Inofita Viotias, north of Athens. With 2700m^2 of collectors, it has an output sufficient to power two adsorption chillers and three compression chillers, each rated at 350kW, which provide air conditioning for the production hall of a cosmetics factory.

Figure 3.46 Solar collectors for a combined heating and cooling installation.

Source: Wagner & Co, Cölbe, from 'Renewables made in Germany' © 2008 Deutsche Energie-Agentur GmbH (dena) German Energy Agency

Figure 3.47 A solar cooling system at the Musée du Bonbon in Uzès, France.

Source: From 'Renewables made in Germany' © 2008 Deutsche Energie-Agentur GmbH (dena) German Energy Agency

Some systems utilize a cold store, where cold water is stored to be used when solar power is not available to power the chiller. Solar cooling works well with a heat pump, as the heat may be returned to the ground or air. The heat pumps can then be used in reverse in the winter, to heat the property.

The chilling of water for drinking purposes is especially widespread all over the world. Small tabletop models are a feature of many offices. But now some small solar thermally driven water chillers are available from several manufacturers and are beginning to make inroads into this huge market. Solar power for the refrigeration of perishables is the ultimate goal for solar cooling. Researchers from the Fraunhofer Institute for Solar Energy Systems (ISE) in Freiburg are demonstrating that this is already feasible, at a winery in Tunisia and a dairy in Morocco. As part of the MEDISCO project (MEDiterranean food and agro

Figure 3.48 A winery in Tunisia, which uses solar refrigeration. The solar collector at the front drives the absorption refrigeration machine. The wine in the fermenter tanks in the background is cooled by a cold accumulator.

Source: © ISE

Box 3.1 Solar thermal heating/cooling system retrofit

Major opportunities exist for active solar thermal systems to be retrofitted to buildings at a time of general refurbishment. The project managers would need to consider the following points, many of which also apply to solar electricity retrofits.

1. Does the timing of the installation fit in with the timing of planned refurbishment work? This applies to both the collectors and the heating system.
2. Have all energy conservation measures being taken prior to sizing the system? Has an energy survey been undertaken?
3. Has research been done to establish the level of available insolation for the location?
4. Has the system been correctly sized?
5. Can the collectors be fitted to the roof or integrated into the facade?
6. Can the collectors be oriented roughly towards the equator? Is there any shading during the main part of the day?
7. Can the collectors be tilted at approximately the optimum angle?
8. Will the building structure accept the additional weight?
9. How will the output from the collectors be integrated into the heating or cooling system?
10. Is there enough space for the heated water storage tank? Is it accessible for installation? Are there any structural changes required for the tank to be installed?
11. Have all the structural and energy system changes to the building been taken into account?
12. Will the building's users be educated in the new system?

Industry applications of Solar COoling technologies) solar plants for refrigerating wine and milk have been installed. The project, funded by the European Commission, is run by the Polytechnic University of Milan. It is only a matter of time before solar refrigeration achieves commercial status.

The future potential

As the often overlooked cousin of PV, solar thermal – despite its much higher installed capacity – has a much greater role to play in the solar revolution. For example, in the US, the primary energy used for water heating, space heating and space cooling comes to nearly 42 per cent of all energy use in buildings, which itself is about 31 per cent of all energy consumption. Building energy use accounts for 38 per cent of US CO_2 emissions and nearly 8 per cent of global emissions. A typical $6m^2$ ($64ft^2$) SWH system produces about the same amount of energy as a 1kW PV system over the course of a year. Utilities, regulators and advocates of renewable energy are starting to understand this, and realize the affordability and value of solar heating and cooling technology. The potential, for job creation, energy security and reducing carbon emissions, not to mention reducing lifetime energy costs, is huge.

Note

1 Figures from the Institute for Sustainable Technologies in Austria.

4

Solar Space Heating

There are many examples of technologies that use the sun's power to heat buildings. However, there are no standard systems or approaches, because each case is specific and demands an individual solution. Space heating systems can be divided into domestic or small scale and district scale; and water- or liquid-based and air-based. We will look at the domestic or small scale first before looking at examples of district heating systems.

Demands of solar space heating

Wherever we live we want heating, particularly in the winter when there is less sunshine. However, we want hot water all the year round. Therefore there is a fundamental design difference between technologies that heat only water for consumption and those that also or only heat space.

It is important from an environmental and cost angle to remember that the goal is not to have a nice piece of technology but to use as little energy as possible. Therefore buildings should be designed for this purpose, and insulated and draught-proofed first. A super-insulated, draught-proofed but properly ventilated passive solar house will require, even in the middle of winter, only a minimum of heating and in some cases (Passivhaus and passive solar building design) none at all except the body heat of the occupants and the heat given off by the electrical appliances in the house, whether it is cookers, fridges or computers. After having gone as far as possible down this route, much of the remaining heating energy demand for water and space could then be provided by the sun or other renewable source.

Small- or domestic-scale systems

These are divided into those which use warmed air as the delivery medium, and those which use water.

Air-based delivery systems

The simplest system is one that just heats air that is drawn directly into rooms by natural ventilation (the stack effect), or by forced ventilation (using fans). These systems are really only appropriate for climates with long, cold winters that have many sunny days. These options may solve specific problems of ventilation or seasonal coldness in a room, perhaps in a commercial or industrial context – a warehouse, storage room or hangar, where it is necessary to keep it above freezing and where the floor or walls have sufficient thermal mass to store the heat overnight. Each system is particular to the location – the building, its orientation and the local climate. Large collector areas are required since there is less sunshine in the winter. This form of space heating represents only 0.8 per cent (1.2GW) of solar thermal applications worldwide. Collectors can be glazed or unglazed. Japan and the US have most of the glazed systems and Switzerland and Canada the most unglazed.

For the maximum comfort of building occupants, the solar heating air should enter the occupied space slowly, continuously and from several different points so that it does not feel draughty. The most efficient operating temperature is around 32°C (90°F) as this is good for comfort and any higher temperature would have a greater heat loss.

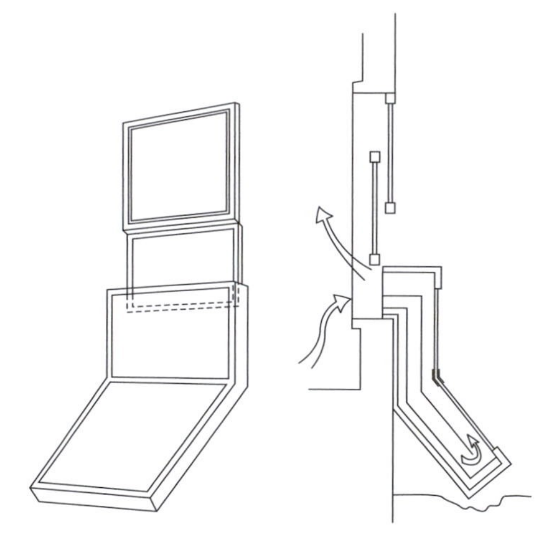

Figure 4.1 The window-box collector.

Source: Kishore, 2009

Window box collectors

This simple system involves having an air-based solar collector beneath a window pointed in the direction of the sun. The collector is a flat, multilayered panel, sealed on three of its four edges and fitted to the outside of the building. Sunlight passes through the glass and strikes the black metal plate, which absorbs the heat. It warms the air flowing around it which rises, and cool air is drawn in at the bottom. The warm air rises into the building beneath the window. The collector's layers are, from front to back: glazing, an upper airspace, a black metal plate, a lower air space, insulation and base. The inlets and outlets are airtight, vented and can be controlled by a thermostat. If the stack effect is not strong enough to draw sufficient air into the bottom from inside, a fan may be used. Thermostatic control of the fan is possible, perhaps powered by a small PV module. The heated air can be directed through ductwork further into the building.

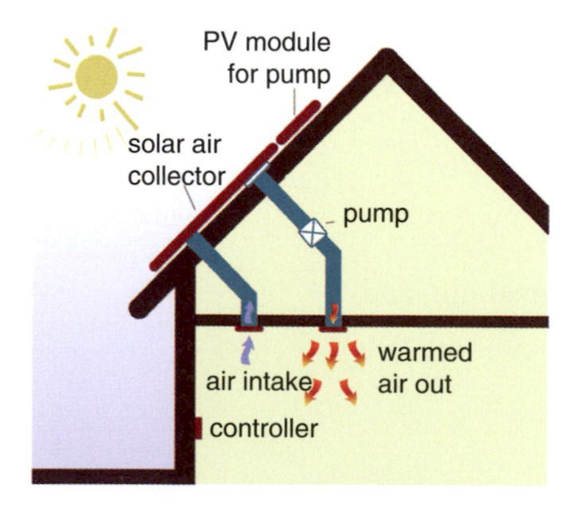

Figure 4.2 An air-based roof-mounted solar heating system: the collector on the roof draws air from within the building, heats it, and returns it to the building. Ideally the warm air is returned at ground level, shown here at ceiling level for simplicity.

Roof- or wall-mounted collectors

These take external air, warm it and use fans to route it into a property. The solar collectors are placed on the roof or equator-facing wall. There are two types of design. One contains material with thousands of tiny capillaries, which absorb the sun's heat. The heated incoming air is then drawn down through ducts into the building where it is discharged around the floor level of the lower floor. Another design is glazed and contains a black plate that heats the air drawn upwards, between it and the glazing. The latter usually draws air from within the building at floor level, as this requires less heating, and releases it at ceiling level.

Rooftop collectors may be connected into a mechanical ventilation system with heat recovery. This ventilation system draws air up through the building into the roofspace via ducts, and removes the heat in it using a heat exchanger, before expelling the used air. The heat is passed to the incoming fresh air drawn from the air-based solar panel. If the incoming air is warmer (on sunny days), the recovered heat is not needed and the heat exchanger is bypassed. But at night and on cold days, the incoming air from the solar panel has the recovered heat transferred to it. This type of system is only suitable for a relatively airtight structure. Because the air's moisture content is normally low when the sun shines, these systems also help to dehumidify the internal climate, displacing internal moist air with warmed dry fresh air, ideal for 'wet' rooms, such as kitchens and bathrooms.

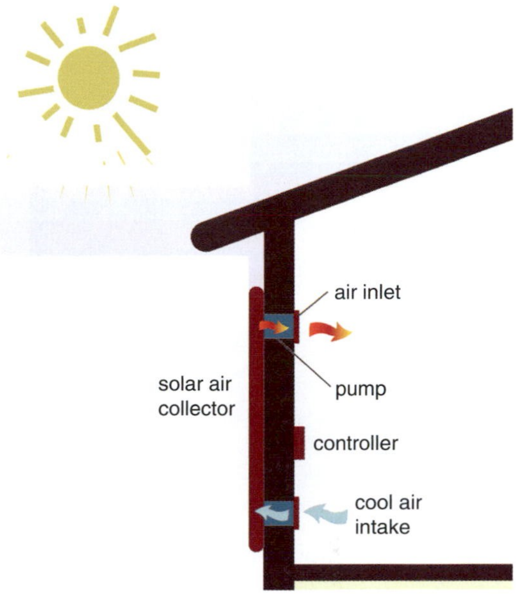

air inlet
solar air collector
pump
controller
cool air intake

Figure 4.3 An air-based wall-mounted solar heating system: the cooler air is taken from ground level and returned, heated, higher up.

The Trombe Wall

Another Frenchman is the inventor of this design – Felix Trombe. Here, the entire face of a building becomes the solar collector and consists of an equator-facing (south in the northern hemisphere; north in the southern hemisphere), glazed, thermally massive, dark coloured wall. The wall is positioned a short distance behind a large area of glazing. At the top and bottom of this wall are ventilation slots through to the building's interior. Cooler air at floor level is drawn by air-pressure differences (the stack effect) into the gap. It rises in the gap between the glazing and the wall, heating up on the way. At the top, it flows back into the room behind it. The design of the airflows within the building can conduct this warmed air into other parts. Additionally, the thermal mass of the wall can also conduct heat from one side to the other and radiate it into the room. The disadvantage of the Trombe Wall is that it removes the view on the equator-facing side of the building.

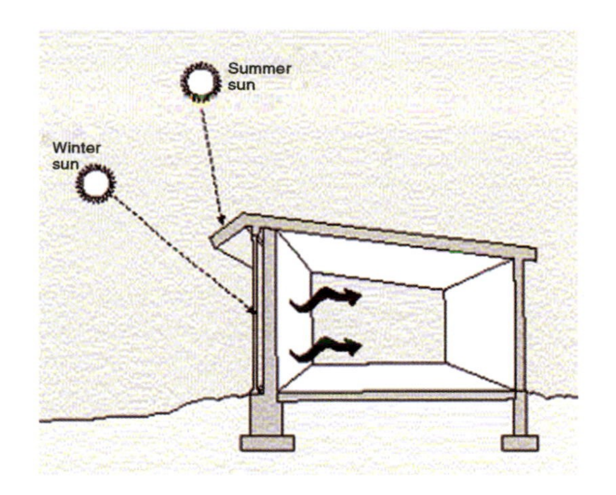

Figure 4.4 A Trombe Wall; the whole of the wall becomes a solar collector.

Source: © Wikimedia Commons

Solar walls

A development of the Trombe Wall, perhaps more suited to industrial buildings, is the solar wall, also called a transpired air collector. An equator-facing wall is clad in aluminium or stainless steel coated with solar absorbing paint. The cladding is a few inches from the wall, and is perforated with thousands of tiny holes. The sun heats up the metal, and a fan at the top of the wall draws up the heated air into a HVAC system. Ideally, this also has built-in heat recovery as above. Ducting transmits the heated air around the building. This is useful for large industrial and commercial buildings, both new and refurbishments. Various companies supply kit solutions.

(a)

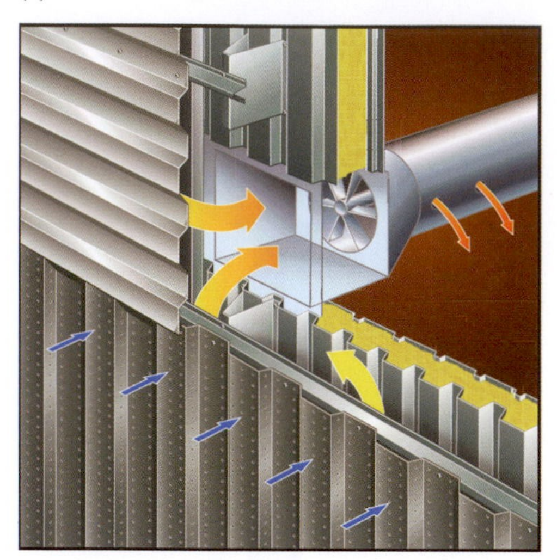

(b)

Figure 4.5 The transpired air collector or solar wall; (a) cutaway diagram and (b) industrial application.

Source: Corus

Water-based delivery systems

Water-based solar space heating systems are commonly hybrid or combi systems, pre-heating the water so that another heat source has less work to do. The auxiliary heat source can be electrical, gas or biomass. Even in Sweden or Canada, it is possible to halve primary energy use with such a system. The chief design challenge is to size it correctly, so that the panel collector area and the storage tank volume match up at the most efficient level. This depends on local weather and latitude, the available roof area, and the pattern of usage of hot water and building occupancy. The larger the solar collector area, the faster the storage tank will heat up; but too large and collected heat will be wasted.

When space heating is required as well as hot water, the energy requirement for a household in a cool climate might be four times higher than hot water alone – around 20,000kWh per year. (The exact amount can be checked for any existing dwelling from its space heating bills, such as gas bills, which normally specify how much gas has been used in kilowatt-hours.) More efficient and smaller homes can use 12,000–15,000kWh, although less efficient houses can consume a lot more. It therefore makes sense to make the home as thermally efficient as possible. Refer to our companion volume *Sustainable Home Refurbishment*, and the chapter on aiming for Passivhaus levels of efficiency.

If an electrical immersion or gas-fired water heater is to be combined with solar space heating, then a smaller storage volume is usually preferred. If a wood-fired boiler, which is not on all the time, is the auxiliary heat source, then provided there is space, a buffer tank should be used for solar-heated water. This makes the most economic use of wood fuel. From an environmental point of view, a biomass-fuelled boiler/furnace is the better option because it has a lower carbon footprint. The solar water heater would supply most of the heat up until winter, and some of it in winter. Biomass fuel would make up the rest of the heating in winter.

A wood burning boiler with a large heat capacity (greater than 20kW) requires a large heat store to hold the heat from the time of production to the time of consumption. Again, the optimum volume for the tank will be decided by looking at the boiler's heat capacity, the heating demand of the occupants and the available space for a second tank. Most well-insulated single-family houses don't actually need the amount of heating that a wood-fired boiler can provide.

A solar space heating system becomes more efficient if the heating is delivered by a radiant heat source. Radiant heating uses a large area of heat emitting surface, such as floors or skirting boards, which only need to be at a maximum of 35°C (95°F) instead of the *c.* 70°C (160°F) for conventional wall-hung radiators. This means that heat pumps can be used in partnership with solar. Ground source heat pumps are preferable to air source heat pumps as they achieve higher efficiencies. Air source heat pumps have significantly reduced efficiency in subzero air temperatures in particular. A solar space heating system will require a buffer tank, whose layout will be similar to that used with a large wood-burning boiler.

How much collector area is required for such a system? In Austria, for example (near southern Europe), the collector area for hybrid solar systems for hot water production and space heating is 12–20m^2 (129–215ft^2) for a single family house. In a well-insulated house, these systems cover about 40 per cent of the total heat demand. However, there are also buildings with considerably larger collector areas and storage capacity. These systems are able to cover

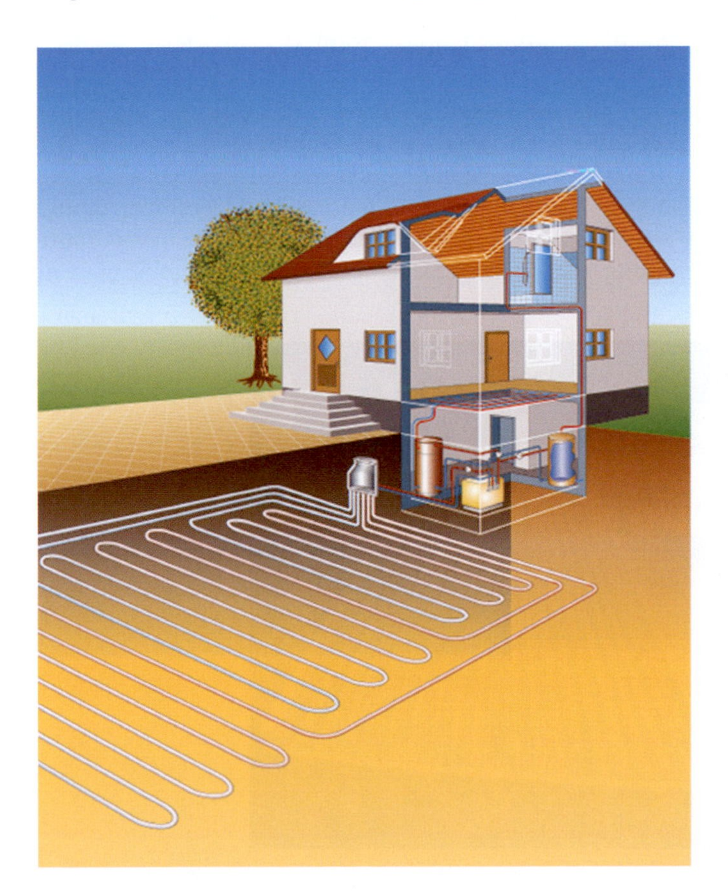

Figure 4.6 Cutaway diagram showing how a ground source heat pump might work in a domestic newbuild, with a cellar and solar water-based space heating system. Note the use of a buffer tank and underfloor heating. The solar collector is located on the equator-facing wall of the house.

Source: Bundesverband Wärmepumpe (BWP), Germany

50–70 per cent of the total heat requirement or even provide 100 per cent solar supply. They will, however, be less efficient and more expensive (per kWh) than systems that meet less of the overall heating demand.

If the property uses an on-demand combi boiler, which has no storage tank, see the section for domestic hot water. There are many installations, particularly in America, Canada, Germany and Scandinavian countries, which use huge tanks for inter-seasonal storage. See the relevant section for this topic.

Case study: John Canivan's system

Some interesting design thought processes relevant to a domestic-scale system are available from a system on Long Island, New York. There, John Canivan built an 8m^2 trickle collector roof-based system, quite stand-alone. He monitored and analysed figures for the middle of winter. He found that, in January, the best days had a maximum sunlight intensity between 600W/m^2 and 700W/m^2. This translates to about 586Wh/m^2 (2000Btu/m^2/hr).

He worked out that this particular collector – mounted at 45° – had a 50 per cent efficiency, and as expected there was a direct relationship between the amount of sunshine, the surface area of the collector and the amount of heat gained. He found a temperature drop of 6°F (3°C) per day from his tank, which only had an R-value of 3m^{2}K/W (= a U-value of 0.33W/m^{2}K, the R-value being the inverse of the U-value). Therefore, in the middle of winter, he needed to ensure there was at least as much heat entering the 200 gallon tank to compensate for the heat going out. For the month of January 2009, the solar heat gain is:

15.5°C × 31 days × 909 litres × 3.6kg/3.79 litres = 87kWh
(60°F × 31 day × 200 gallons × 8 lb/gal = 297,600Btu)

According to figures from NASA (see Figure 4.9), January and December should yield about 2–3kW/m^2 on Long Island. John found that, in fact; it was more like 1kW at his site during that year. This emphasizes the significance of

Figure 4.7 John Canivan's system, Long Island, US.

Source: John Canivan

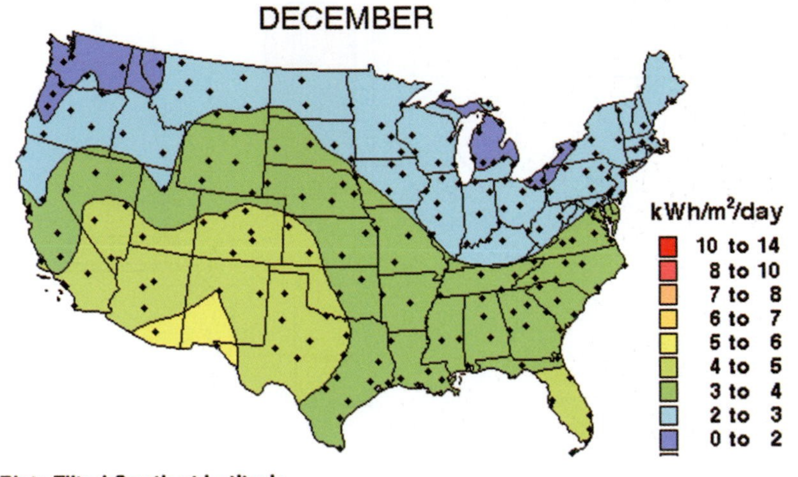

Figure 4.8 Insolation map for the US in December.

Source: NASA

location – there was some shading and clouding due to local landscape morphology. Despite this, he logged a 88kWh (300,000Btu) heat gain to a partially shaded array. 'In the month of January,' he recorded, 'this may not seem like much but with an array four times larger (which I recommend) the heat gain for the month of January on Long Island could easily be 586kWh (2,000,000Btu).

Case study: A passive and active solar house in Wisconsin, US

Daycreek is a house belonging to a couple known as Jo and Alan Stankevitz in South East Minnesota. The two-storey house is a post and beam frame, heated using passive and active solar systems, a wood stove and electric boiler as a back-up. Electricity is supplied from the local grid, and a 4.2kW PV system. On sunny days during winter, the bulk of heating needs are supplied by both passive and active solar heating systems. Passive solar is provided by the sunlight that streams into the south-facing windows shaded in the summer by 1m (3ft) overhangs.

The active solar heating system consists of ten 1.2 × 3m (4 × 10ft) solar collectors, two 24V, 75W pumps powered by two 75W, Siemens PV panels and a 364 litre (80 gallon) domestic hot water tank. It was built to heat an insulated sand bed beneath the floor of the house. Because it took about 12 hours for the heat to travel through the sand bed into the floor, it was discontinued. A larger water tank is recommended instead. When the sun doesn't shine, the house is heated by a 18kW (60,000Btu/hr) wood-burning soapstone stove. As a back-up, there is a 4.5kW (30,000Btu/hr) electric boiler.

Industrial-scale solar thermal power

In 2008, a survey found that solar thermal plants in industry form a very small fraction (0.02 per cent) of the total solar thermal capacity installed worldwide, which equals 118GWth. Only some of this solar heat will be for space heating, and the rest is destined for process heat. The amount of heat required for an industrial building depends on the level of insulation, the compactness or

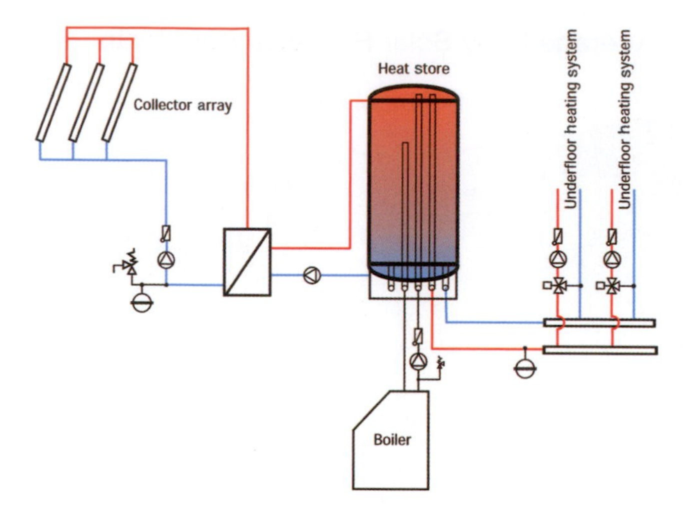

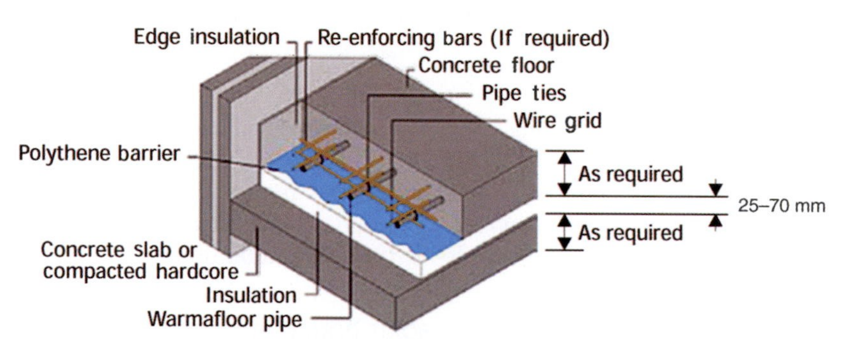

Figure 4.9 Typical concept diagrams for an industrial-scale system, as used in hangars and factories. An array of solar collectors on the roof feeds an optional, suitably sized storage tank supplemented by a boiler. This serves an underfloor heating system, where the pipes are laid in tandem with a reinforced concrete wire grid. The instance in the photograph is of an aircraft hangar.

Source: IEA-SHC

Figure 4.10 Solar collectors for space heating on a Passivhaus standard office building for Neudorfer, in Rutzemos, Austria.

Source: © IEA-SHC

profile of the building and its use. Internal gains, infiltration and ventilation losses are greatly influenced by heat use. Most industrial buildings are of a lightweight metal construction for the walls and roof but with a heavy concrete floor. Preferably, the walls and roof contain an insulation layer in between outer and inner sheet metal layers. A solar-heated building is mainly designed for passive heating and cooling, but thermal modelling is required to determine the degree of additional heat required for the operating conditions. Figure 4.9 shows a typical schematic layout. Sometimes it is possible to dispense with a thermal storage tank if it is deemed that the thermal mass of the concrete floor is sufficient, once heated, to maintain a reasonable and consistent temperature throughout the year, day and night. An insulation layer with a minimum depth of 23cm (more than 9 inches), should be installed underneath the concrete base, and thermal bridging minimized throughout.

One example is a warehouse with integrated office for Neudorfer, in Rutzemos, Austria, built in 2005 (Figure 4.10). The section of the building containing the offices is built according to Passivhaus standards with an annual space heating consumption of 18kWh/m^2. Solar thermal collectors installed on the roof feed into an underfloor heating system, while facade-integrated PV panels supply electricity.

Process heat

It is worth mentioning here the use of solar thermal for process heat in the industrial sector. This sector accounts for 30 per cent of energy demand in OECD countries. Of this, two-thirds is for heat and one-third for electricity. Of the heat demand, almost three-quarters is in the low (up to 100°C [212°F]) and medium (up to 400°C [752°F]) temperature range, which can be satisfied by solar thermal technology. With only around 100 solar thermal plants for process heat known in the world, the potential is large; considered to be up to 3 per cent of total world energy demand. The key sectors that could make use of such heat are food, including drinks, textiles, transport equipment, metal and plastic treatment, and chemical. It could be used for cleaning, drying, evaporating and distillation, pasteurization, sterilization, and more, besides solar space heating and cooling. The areas of the world that could benefit the most from exploring this source of process heat are those in the arid and hot, Mediterranean and tropical regions. The full range of collectors, including all types of parabolic dishes and troughs, described in the chapter on solar electricity, can be used to serve process heat; the technologies are available for different applications. The best independent source of information on this topic is the IEA, which has published the results of its research in this area, as part of its excellent solar heating and cooling programme.

District solar space heating

District heating systems using solar energy have been in operation since the 1970s. They cannot supply all of a district's heating needs throughout the year, but they can make a significant contribution and improve security of supply. Another benefit is that operational costs are fixed and predictable and, of course, they contribute to the reduction of greenhouse gas emissions. They make an ideal partner to combined heat and power (CHP) systems. Pioneering countries in this technology include Sweden, Denmark, Germany and Austria. With 20 years' operational experience of such plants to draw on, there is now interest in the commercial operation of solar district heating from utility companies, local authorities and the housing sector.

Solar district heating plants collect their heat from large collector fields and often store this heat in large underground storage tanks. In some cases, these use natural underground caves or reservoirs, in others they are specially excavated, highly insulated chambers. Distribution of the heat is via heating networks to residential and industrial buildings. District heating plants of this nature often have supplementary heat supplies, which can be integrated in series or in parallel. Storage can also be decentralized, with each dwelling or business unit having its own storage tank. In this case, supplementary heating takes the form of a boiler in each unit. Most systems in southern Europe are used for supplying hot water, and in northern Europe for space heating. Those in northern Europe most commonly have central heat storage and supplementary heating provision. Technical solutions are available to meet a wide range of needs. In centralized systems, protection against *Legionella* is ensured by heating the much smaller volume standby tank each day to 60°C (140°F).

Heat can be distributed via the collector circuit, using heat transfer units in individual apartments, or centrally via circulating pipes. Collectors are prevented from freezing in winter by adding glycol, although drain-back systems are used in the Netherlands (see Chapter 3). Control and safety features protect systems from overheating in the summer. When used to supply hot water, these systems generally meet up to 60 per cent of heat demand.

The successful design of such systems requires an accurate estimate of hot water needs. If actual consumption is below or over the estimate, the system will be inefficient. They can be easily integrated into existing central heating and hot water systems. In Europe, the current estimate of cost of the system, including the pipework, storage unit, control system and design, is €600–€1200 per square

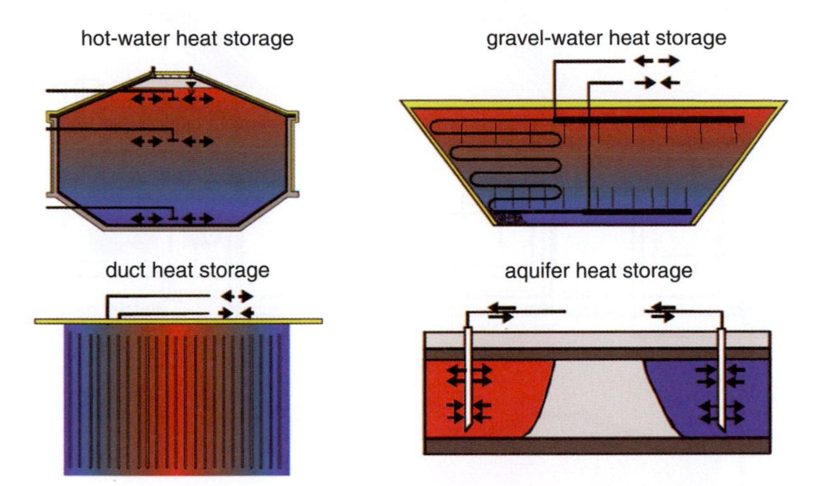

Figure 4.11 Various types of interseasonal heat stores for larger-scale heating networks.

Source: SOLITES

metre of collector area. Variables include the collector type, the design philosophy, and the situation of the buildings. Partly, of course, it also depends on the percentage of the heating demand that is to be satisfied by solar power. To give a rough idea, in central Europe, for example, to meet half of the annual heating needs for hot water requires a collector area of 1–1.5m^2 per apartment and about 50–60 litres (11–13 gallons) of solar heat storage per square metre of collector surface. If supplementary heating is required too, then roughly 3–4m^2 (32–43ft^2) of flat plate collector per apartment is needed. This will result in 15–20 per cent of the heating coming from the solar system. Complete solar district heating networks require 10–30m^2 of flat-plate collectors per apartment to yield half of the annual energy requirements. This equates to around 1.5–2.5m^2 (16–27ft^2) of collector area and up to 5m^3 of storage volume per 1000kWh of annual heat demand. Additional savings can be made by integrating the flat-plate collectors with a facade or roof of a building.

Hot water storage tanks are highly insulated containers. For gravel/water heat storage, the pits are lined with special sheeting and filled with a water/gravel mixture. The heat is transported to the store by the water or through coiled pipes. This system has a lower heat capacity than straight hot water storage – and so requires about half as much volume again. With borehole or duct storage, solar heat is transferred to the ground via boreholes 20–100m (65–328ft) deep. Due to the lower thermal capacity of soil, the stores are generally three to five times larger than hot water units, but are less complicated to build and can be extended as needed. They function best when there is no direct contact with groundwater. For aquifer heat storage, groundwater is used to store the energy. It is raised from the borehole, warmed by a heat exchanger and returned to the aquifer via a second borehole. Heat is recovered by reversing the flow. The cost per heat unit is low but, of course, this does require specific hydrological conditions.

Case study: Friedrichshafen district heating system

One example of a district heating system, operational since 1996 and servicing 600 homes, can be found in Friedrichshafen, Germany. The city is in south-western

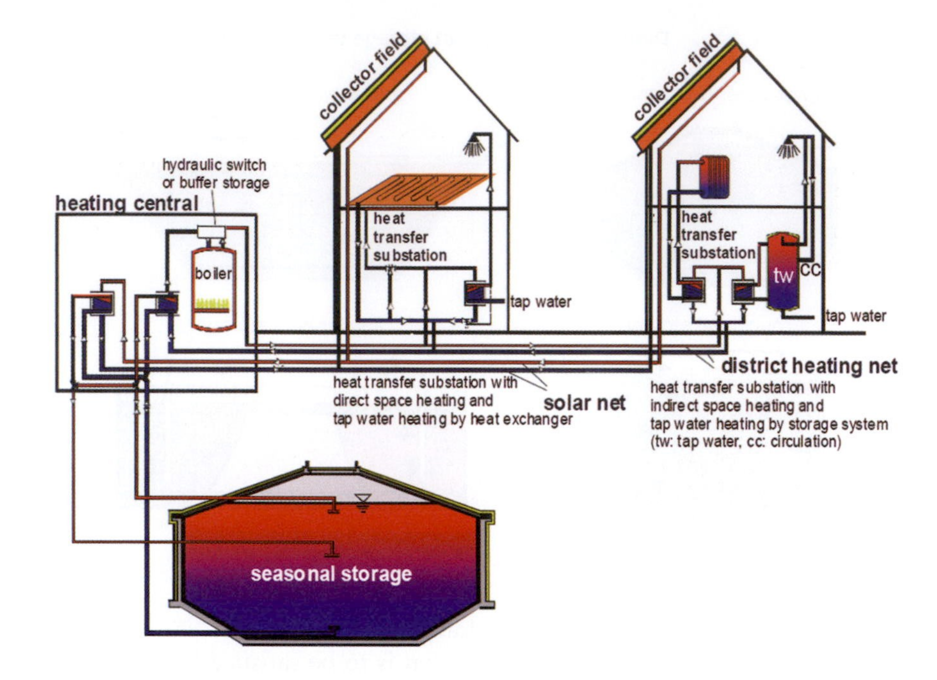

Figure 4.12
Friedrichshafen district heating system, Germany.

Source: IEA-SHC

Germany and has an annual mean temperature of 9.7°C (49.5°F), and its degree days figure (see Resources for an explanation), base 17°C (65°F), is 3717. The homes were built in the 1990s with a heating system planned from the outset. A total of 4300m² of solar collectors were installed on the roofs of multi-storey buildings. The solar heat collected in the summer is fed into an underground storage network at temperatures of 40°C–90°C (104–194°F). This heat is discharged into the distribution network by another heat exchanger. The heating and hot water in each home are themselves separated off by a heat exchanger in separate heat transfer stations. The goal of the pilot project was to cover almost half the total demand for space heating and hot water supply. The other half is supplied by gas-fired condensing boilers.

Some of the collectors are mounted and tilted on the roof, while others are integrated into roofs. The heat storage tank is cylindrical and made of reinforced concrete lagged with 20–30cm (8–12 inches) of Rockwool, underlined with stainless steel sheeting as protection against vapour diffusion. This pilot project underperformed compared to expectations. More recently the system has been expanded and improved based on this knowledge.

Case study: The Drake Landing Solar Community

The Drake Landing Solar Community (DLSC) is a settlement in Okotoks, Alberta, Canada, equipped with a central solar heating system and other energy efficient technology. It was the first of its kind in North America. There are 52 homes heated by 800 solar thermal panels on four rows of garages, with two rows of collectors per garage. They generate 1.5MW of thermal power during a typical summer day, which is fed into an underground interseasonal heat store. There are also two large, horizontal, insulated water tanks 3.66m (12ft) in diameter and 11m (36ft) long. The water temperatures inside are stratified to improve the overall efficiency. From there, plastic, insulated pipes distribute the heated water at 35–50°C (95°–122°F), depending on the outside air temperature. The flow is regulated to match demand. The relatively low temperature reduces losses from the pipes and helps the solar collectors operate more efficiently. To meet any shortfall from the solar system, each home is equipped with a specially designed air-handler unit for adequate heat distribution with a user-operated thermostat for individual comfort. The project hopes to receive over 90 per cent of its heating needs from the system by its fifth year.

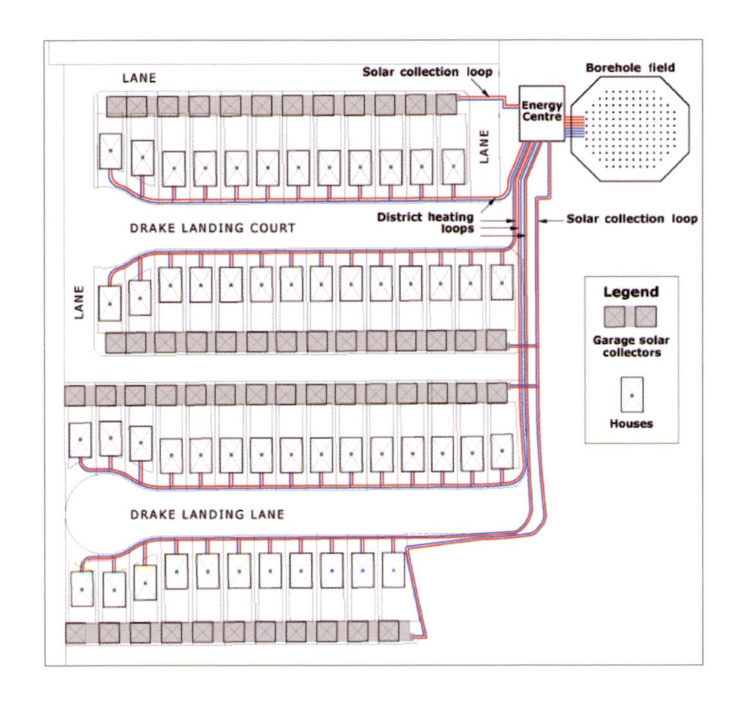

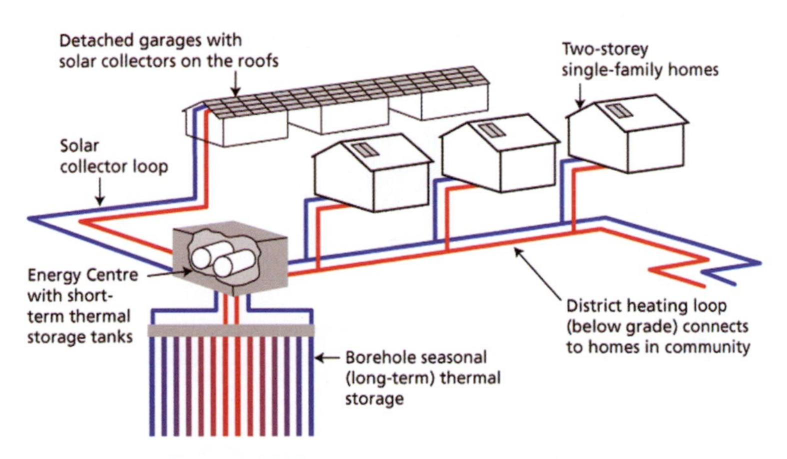

Figure 4.13 Plan and schematic diagram of the district heating system in Okotoks, Alberta, Canada.

Source: Drake Landing Solar Community

During the warmer months, the heated water is distributed from the short-term storage tank to the borehole thermal energy storage (BTES) system via a series of pipes. The pipes run through a collection of 144 holes that stretch 37m (121ft) below the ground and cover an area 35m in diameter. As the heated water travels through the pipe-work, heat is transferred to the surrounding earth. The temperature of the earth will reach 80°C (176°F) by the end of each summer. To keep the heat in, the BTES is covered with sand, high-density R-40 insulation, a waterproof membrane, clay and other landscaping materials. The water completes its circuit of the borehole system and returns to the short-term storage tanks to be heated again and repeat the same process.

The location is 51.1° N, 114° W, with a height above sea level of 1084m (3556ft). The temperature varies from –33°C (–27°F) in the winter to 28°C (82°F) in the summer. The azimuth for the panels is south with a tilt of 45°. The panels are fed through a closed loop containing 50 per cent propylene glycol antifreeze. This solar heat is transferred to the heat exchanger from where it goes to the water in the short-term storage tanks. The flow rate through the collectors is constant, but that on the water side of the heat exchanger is automatically adjustable, allowing the control system to set a desired temperature rise.

The system is constantly monitored and the data published in real time online at www.dlsc.ca. The illustration in Figure 4.14 is a snapshot taken at the date and time caught in the image – early in the morning in June.

Solar heating and cooling-combined heat and power systems

Medium- to large-scale solar heating and cooling systems, used for district or commercial applications, will typically need supplementing with another energy source. CHP from a gas-powered generator is attractive for this purpose because of its efficiency. The next most sustainable alternative would be a biomass boiler, but this would not generate electricity. Such a hybrid system

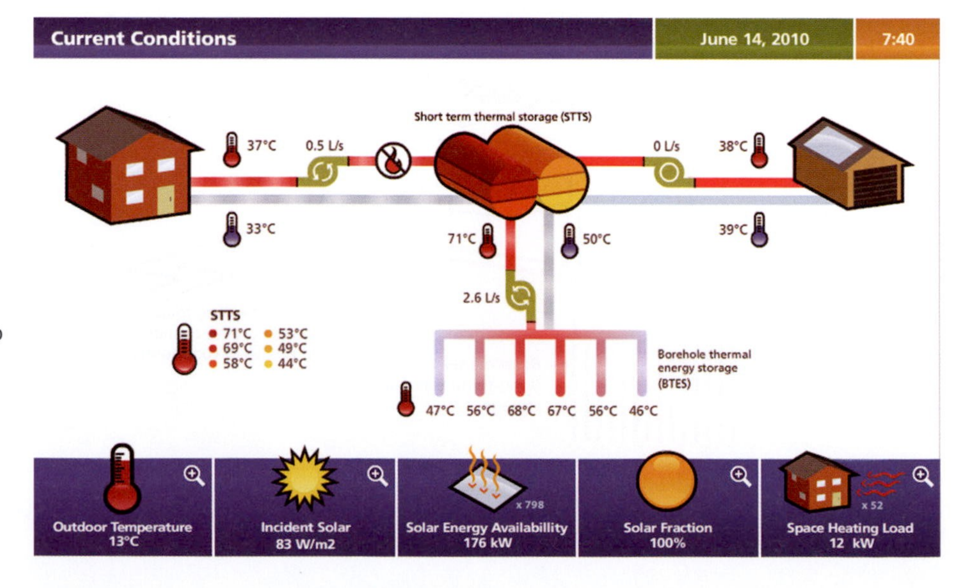

Figure 4.14 Screen grab from the Drake Landing Solar Community website, www.dlsc.ca, showing live readings of their system performance.

Source: Drake Landing Solar Community

uses an absorption chiller for the cooling function, using heat recovered from the CHP system, which not only meets the summer cooling load but reduces the peak electric demand caused by the cooling requirement. In addition, a balanced heat demand over the year, due to the presence of the absorption chiller, improves the economics of the whole process by increasing the number of hours per year it can be operated. There are great efficiency savings, especially in the cooling season, as the heat from the CHP unit can be used for the heat-driven chillers rather than going to waste.

The main challenge for a hybrid system involving CHP is to match the two components, because of the radically different behaviours of the solar collectors and the co-generation plant – the irregular heat supply of the solar component with the regular operation of the CHP – where the main purpose of heat recovery is really to cool the engine. A resource to help design engineers address this, and find the optimum match for the two different components, has been developed by the IEA: Napolitano, A., Franchini, G., Nurzia, G. and Sparber, W. (2008) 'Coupling solar collectors and co-generation units in solar assisted heating and cooling systems', Eurosun, Italy (found along with other useful resources at www.iea-shc.org/publications/task.aspx?Task=38, accessed 6 February 2011).

5

Further Solar Thermal Power Applications

Solar thermal power has an astonishing array of applications other than for water and space heating and cooling. In this section, we will examine ways to use the free energy from the sun to help with desalination, humidifying, growing, drying, sterilizing, pasteurizing and cooking.

Desalination and distillation

Where polluted water needs to be purified in semi-tropical and tropical climates, or fresh water made from salt water, the power of the sun can be brought into play. A variety of techniques are available.

Solar stills

A basic solar still will remove impurities by evaporating and condensing the water and leaving the impurities behind. The distilled water can then be safely drunk or used in hospitals, or in cooking. Thousands of these small-scale stills are in operation around the world.

A simple still is easily made using glass or transparent plastic as a cover, placed at an angle over a shallow tray of water. Glass is better than plastic, which tends to degrade in ultraviolet light and doesn't allow water to condense so easily onto it. The tray has a black backing to trap as much as possible of the sun's energy. As the sun heats the water, it evaporates and condenses on the underside of the glass or plastic and drains down the slope, where it is collected in a tank. The slope should ideally be at an angle of 10–20°.

The higher the initial temperature of the water and the cooler the condensing surface, the more efficiently the still will operate. The following will maximize this efficiency:

- low-E glazing;
- a tray for the water, which is insulated on all sides except the one facing the sun;
- a black surface;
- water depth of a shallow 20mm (0.79"), so it heats more quickly;
- the collector tilted at an angle greater than 15°, ideally perpendicular to the midday sun, facing south in the northern hemisphere and north in the southern hemisphere;
- the collector situated away from the shadows of buildings or trees.

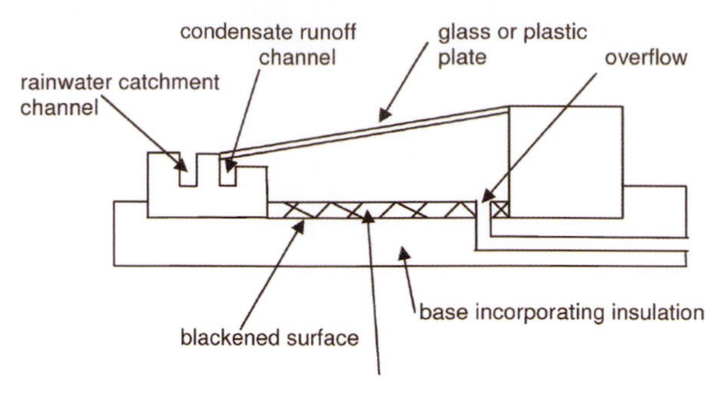

Figure 5.1 Schematic of a basic solar still.

Source: Otto Ruskulis for Practical Action

The efficiency of solar stills can vary between a maximum of 50 per cent and a typical one of 25 per cent. They work best in the early evening when the feed water is still hot, but the air temperature is falling. A cool wind will also help to keep the condensing surface at a lower temperature. Problems can occur due to poor maintenance, where fittings develop gaps that let in cold air. Cracking, dirt, algal growth or scratches on the glass or plastic reduces the solar transmission. The tray must also be regularly cleaned of the residue and not allowed to dry out when in use. The still can also be employed to gather rainwater by adding an external gutter to catch the run-off.

A rough figure for the amount of potable water that can be produced, in an area where the global daily solar irradiation is around 5kWh/m^2, assuming that the still has an operational efficiency of around 30 per cent, is about 2.3 litres per square metre per day. Averaged out over a year, this might be one cubic metre of water per square metre of still. People need about 5 litres (1 gallon) of drinking water a day. Therefore, about 2m^2 (22ft^2) per person is the minimum area requirement. If significantly more water is required for the available surface area, then a PV solution may be preferable. This uses PV-powered reverse osmosis. For more information on this subject, see page 175. Note that it takes more energy to distil water than it does to pump it, using PV power. Therefore, if a supply of fresh water is nearby that can be pumped, it may be more cost-efficient to use this solution.

Distillation is only appropriate to remove dissolved salts. If the problem with the polluted water is merely suspended solids, then a sand filter may be a better option.

Water-cooled greenhouse

Evaporative cooling (see the section on passive solar architecture) and desalination can be combined in hot arid regions by the sea, to create an optimum space for crop growing and provide clean water.

The greenhouse is constructed next to a body of water, which may be seawater. The water is drawn onto cardboard spread over the ground along one side of the greenhouse. It evaporates and the humidified air is drawn into the greenhouse. This provides a cooling effect that reduces the temperature within the greenhouse. At the other end of the greenhouse, a condenser captures the clean water from the air. The condenser uses cold water as a coolant. The greenhouse is made of a light steel structure with a polythene or glass covering. The polythene films are treated to incorporate ultraviolet-reflecting and infrared-absorbing properties. The cardboard evaporators become strengthened by the crystallized calcium carbonate from the water (Figure 5.2).

There is one project that is taking this idea on to a much larger scale. The Sahara Forest Project aims to develop an integrated, large-scale system for the

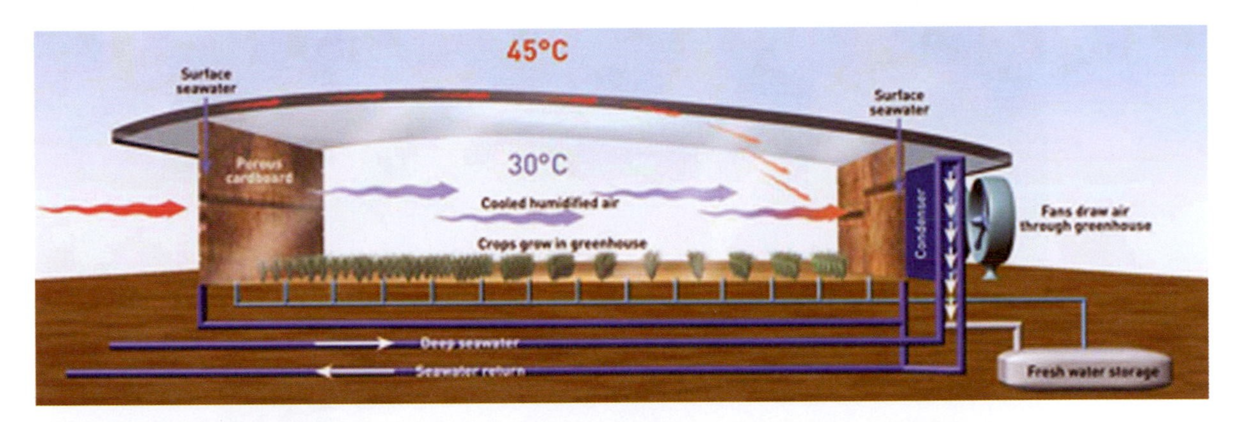

Figure 5.2 The principle of this seawater greenhouse is that seawater is evaporated by the intense solar heat at the front of the greenhouse to create cool humid conditions inside. A proportion of the evaporated seawater is then condensed as fresh water that can be used to irrigate the crops. The air entering the greenhouse is first cooled and humidified by seawater, which trickles over the first evaporator. This provides good climate conditions for the crops. As the air leaves the growing area, it passes through the second evaporator over which seawater is flowing. This seawater has been heated by the sun in a network of pipes above the growing area, making the air much hotter and more humid. It then meets a series of vertical pipes through which cool seawater passes. When the hot humid air meets the cool surfaces, fresh water condenses as droplets that run down to the base, where they can be collected.

Source: © Seawater Greenhouse Ltd

production of fresh water, energy and crops. A Sahara Forest Demonstration Centre is under way. In addition, the world's first commercial Seawater Greenhouse is already operational in South Australia (Figure 5.3).

The cool and humid conditions in the greenhouse enable crops to grow with very little water. When crops are not stressed by excessive transpiration, both the yield and the quality are higher. The process imitates the hydrological cycle where seawater heated by the sun evaporates, cools to form clouds, and returns to the earth as rain, fog or dew.

Heat pumps for cooling

Using heat pumps for cooling is a valuable process in areas where the climate induces large seasonal temperature differences between summer and winter as well as between day and night. The principle of a heat pump is the same as for a fridge, but in reverse. Imagine that heat, instead of being pumped out of the fridge, was being pumped into it from the outside. Low-temperature air from a large volume would be concentrated in a smaller volume, raising its temperature. If, instead of a fridge, we have a building or a room, with the heat being pumped in from outside, then the same principle applies on a larger scale. Heat pumps can take heat from, or return heat to, the ground, air or a nearby body of water if it is available.

Heat pumps are powered by electricity and are judged by their COP. This is the ratio of the amount of heat or coolth produced divided by the electricity consumption of the heat pump. So, for example, a heat pump with a COP of 3 (or 3:1) will produce three times as much heating energy as the electrical energy it consumes. The higher the CoP the better the performance.

Figure 5.3 The Seawater Greenhouse, Port Augusta, South Australia, which began producing water, vegetables and sea salt in autumn 2010, just before the fittings were installed.

Source: © Seawater Greenhouse Ltd.

Many heat pumps are reversible, and can be used for cooling. A valve can change the direction of the refrigerant flow to cool the building. Even with water-to-water heat pumps designed for heating only, a limited amount of 'passive' summer cooling can be provided by direct use of the ground loop, for example, by means of bypassing the heat pump and circulating fluid from the ground coil through a fan convector. Carbon savings are maximized if the electricity supplied is from a renewable source like a PV array.

Solar drying

All over the world, drying is the commonest method of preserving food. The simplest form of drying is just leaving the product – whether it is tomatoes, spices, herbs, nuts, peppers, beans or fish – out in the sun on mats, roofs or floors. Of course, the advantage is that it is almost free and can be done close to the home. But if you want to ensure that the food is not contaminated by dust or spores, then solar dryers are a simple if slightly more expensive solution.

Solar dryers have other advantages too: as they reach higher temperatures, the drying time is shorter and more of the moisture is removed. More food may be treated in the same amount of space and kept protected from contamination and rain. A solar dryer will work even when the sun isn't shining. Use of solar drying for large-scale operations will often save on the firewood or electricity, which would otherwise have been used.

Basic design

The basic design of a solar dryer consists of an insulated compartment or box, with a plastic or glass cover. Warm air is drawn into the bottom and rises, absorbing moisture on the way. It is allowed to leave near the top, taking the moisture with it. The size and shape of the box depends on the products to be dried and the scale of the drying system. Large systems can even use barns, while smaller systems may just hold a few trays stacked one above the other in a small wooden box.

The inside of the collector is, as with the solar still, painted black to absorb the sun's heat. It is best to use glass rather than polythene. The collector must be tilted towards the sun at an angle greater than 15° to allow rainwater to run off. Again, it should be perpendicular to the midday sun, facing south in the northern hemisphere and north in the southern hemisphere, and situated away from the shadows of buildings or trees.

Tent solar dryer

A tent solar dryer is another basic form. It consists of a ridge pole over which plastic has been draped and fastened down at the base. Black plastic should be used on the wall facing away from the sun. The food to be dried is placed on a rack above the ground. The ends of the tent are blocked off to prevent dust from entering and to keep in the heat. This form of dryer is vulnerable to winds.

Brace solar dryer

The Brace solar dryer is a wooden box with a hinged transparent lid (Figure 5.4). The inside is painted black or lined with black plastic and the base is insulated. The food is positioned on a mesh tray or rack above the floor. Air is allowed to flow into the chamber through holes at the front and leave through vents at the top and back.

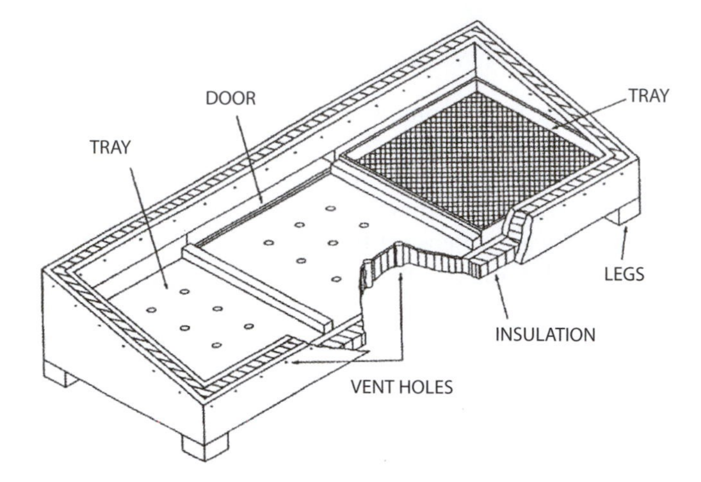

Figure 5.4 Schematic diagram of a Brace solar dryer.

Source: Practical Action, Schumacher Centre for Technology and Development

Indirect solar dryer

A more sophisticated version has a solar collector at ground level directing the hot air into a vertical cupboard containing mesh shelves or racks that hold the food. The air leaves through a chimney or flue at the top. This has the advantage that the food is not in direct sunlight. Heated air passes all around it, drawn up by the stack effect (Figure 5.5). This is called an indirect solar dryer.

Solar kiln

A solar kiln has been developed for seasoning timber. It is similar to a greenhouse (although only the sun-facing sides need to be glazed), but with air drawn in at the base and allowed to exit at the top, where there is an opening protected from rain infiltration. Wood can be stacked inside to allow the hot air to circulate around it. The kiln dramatically speeds up the seasoning process.

Solar pasteurization

A similar system can also be used for solar pasteurization. Millions of people become sick each year from drinking contaminated water. Pasteurization involves purifying water by heating it to 65°C (149°F) for about six minutes to remove pathogens. It is not necessary to boil it. This will kill all germs, viruses and parasites that are carried in water or milk and cause disease in humans, including cholera and hepatitis A and B. Milk is commonly pasteurized at 71°C (160°F) for 15 seconds.

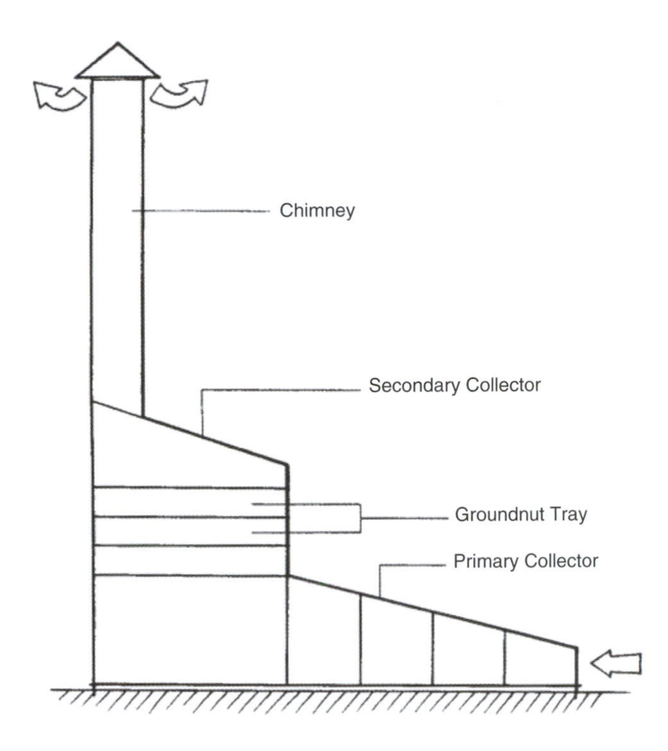

Figure 5.5 Schematic diagram of an indirect solar dryer.

Source: Practical Action, Schumacher Centre for Technology and Development

Water or milk can be pasteurized in a solar cooker. The pioneers in this area are Dr Bob Metcalf and his student David Ciochetti. In the early 1980s, they found that when contaminated water was heated in a black jar in a solar box cooker, both bacteria and rotaviruses – the main cause of severe diarrhoea in children – were inactivated by 60°C (140°F). In 1984, it was concluded that if

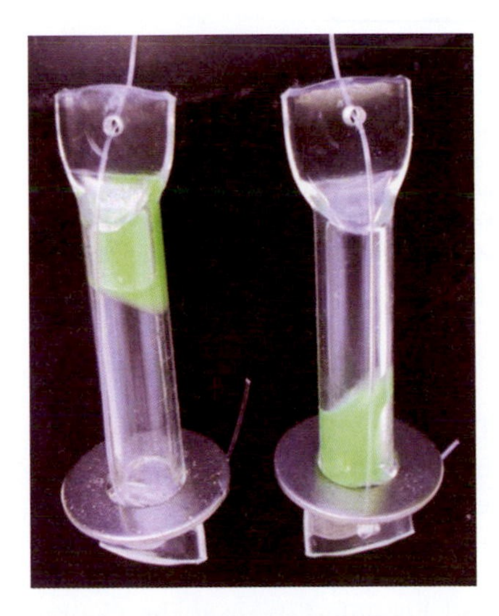

Figure 5.6 Solar Cookers International's WAPIs shown in the solid wax state (left) and liquid wax state (right).

Figure 5.7 Boniphace Luhende in Tanzania demonstrates solar water pasteurization using a CooKit (a brand of solar cooker) and WAPI.

contaminated water were heated to 65°C (149°F), all pathogenic microbes would be inactivated.

Since thermometers are not accessible to many people around the world, there is a need for a simple device that indicates when water has reached pasteurization temperatures. In 1992, Dale Andreatta (a mechanical engineering researcher at the University of California), developing a 1988 design by Dr Fred Barrett, created the Water Pasteurization Indicator (WAPI) in its current form (Figure 5.6). The WAPI is a clear plastic tube partially filled with a soybean wax that melts at about 70°C (158°F). With the solid wax at the top end of the tube, the WAPI is placed in the bottom of a black container of water that is solar heated. If the wax melts and falls to the bottom of the tube, it indicates that water pasteurization conditions have been reached. The WAPI has a stainless steel washer around it to keep it at the bottom of the container, which is the coolest location when solar heating water. An automotive thermostat can be used to control the flow of water through a flow-through water pasteurization device.

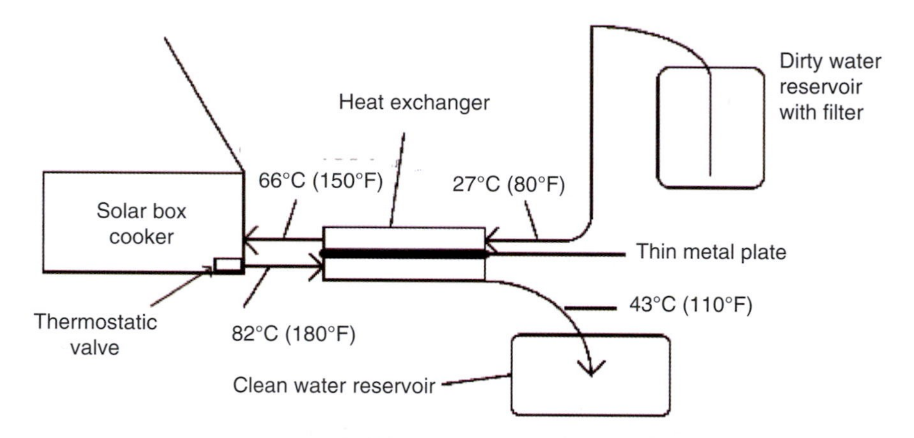

Figure 5.8 Schematic diagram of the solar water pasteurization device. The dirty water is fed across a thin metal plate into the solar box cooker. On the way, it takes heat from the clean water that has been heated in the solar box cooker. A thermostatic valve only allows the water to leave the cooker when it has reached the required temperature. On the way to its tank, it passes again through the heat exchanger, which cools it, passing some of its heat to the incoming water. Since top water temperatures are often 2–5°C (36–41°F) hotter than bottom water temperatures, lower WAPI placement helps to further ensure that pasteurization conditions have been achieved.

Solar cooking

How about a meal cooked by sunshine? There's no space in this book for recipes but there is a wide choice: food can be boiled, fried, roasted and baked using solar cookers. They are used just like a standard stove to sauté or fry foods, or even bake bread – not to mention boil water for coffee or tea.

Solar cookers were invented by American astrophysicist Charles Greeley Abbot in 1915 and one was used by him and his wife Lillian at the George Ellery Hale's Mount Wilson Solar Observatory to bake and roast and to feed the local staff. Abbot was a solar pioneer, and built many devices for measuring solar radiation.

Design and principles

Using one of these devices, the dish of the day would more than likely be cooked over a dish, since it is often a parabolic reflector – frequently called a 'dish' – which is used for this purpose. When aimed at the sun, all the light and heat that falls upon its mirrored surface is reflected to a point known as 'the focus'. If a black cooking vessel, such as a casserole dish or a frying pan, is placed at the focus, it will absorb the energy and become piping hot.

There are other designs of solar cooker, such as boxes and troughs. All incorporate brightly reflective fins or surfaces. Smaller versions can be folded up to be easily portable. Primitive versions can be constructed using cardboard and silver foil. A satellite dish can even be adapted into a cooker. More sophisticated models comprise sheets of anodized aluminium, or they can simply be bought off the shelf.

Whatever the design, the principle is always the same: to concentrate the sun's heat onto the pot. Consequently, the cooker must be pointed squarely at the sun, and, usually, moved manually and frequently, to follow the sun around the sky during the cooking period. Only direct sunshine is appropriate. The larger the collector and the tighter the focus, then the higher the temperature will be at the focal point for a given level of insolation. Temperatures at the pot can soon cause it to boil and continue to simmer. The cooking temperature can be adjusted by deliberately misaligning the collector if required. The technique is suitable anywhere where the daily direct sunshine reaches a temperature of over 24°C (75°F).

Figure 5.9 A parabolic dish cooker with the cooking pot at the focal point.

Source: Creative Commons

Care should be taken when tending the pot because it can get extremely hot where the Sun's rays are focused.

Solar-cooking applications

Solar cookers are, for one thing, an alternative to using a barbecue – whether while camping or at home – which saves the trees used to make charcoal. They are also valuable in parts of the world where cooking is normally done on a fire. It is quite common, for example in Africa or India, for rice, bean or chicken stews to be cooked in this way and left on the fire for hours, which is not necessarily a good idea since too long a period of cooking destroys nutrients. Much deforestation occurs around the world from the need for firewood for cooking, and a lot of people's time is spent sourcing this fuel. There are also considerable health risks from the daily inhalation of fire smoke. Therefore, solar cooking not only reduces firewood needs, but saves health, time – or money if the wood fuel is bought – and combats air pollution.

The disadvantages of solar cooking are that it can only be done when the sun is shining and the cooking time is often much slower than conventional cooking (allow at least twice the time). The practice must therefore be adapted to suit. For these reasons, solar cooking can never completely replace conventional cooking.

Projects to introduce solar cookers to a region usually involve some form of training. Pupils are taught how to cook local traditional dishes with the new equipment by adapting the recipes and cooking methods and times to suit the solar cookers. They also learn how to clean the cooker to keep its surface shiny. For example, in the Dominican Republic, a project to introduce solar cookers has a condition that before a family can purchase or earn one, they must learn to use it and cook several meals at the community centre.

The 'hot box'

Solar cookers can be used in conjunction with a 'hot box', or 'hay box' (Figure 5.11). This is simply an insulated box into which a pot that has been

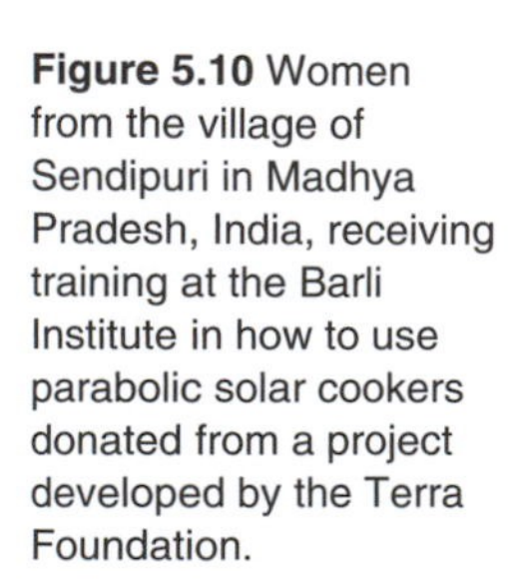

Figure 5.10 Women from the village of Sendipuri in Madhya Pradesh, India, receiving training at the Barli Institute in how to use parabolic solar cookers donated from a project developed by the Terra Foundation.

Source: Barli Institute

brought to the boil can be inserted. It will then carry on simmering for long enough to cook the meal. They can be bought or home-made, in which case they may be insulated with old polystyrene packaging, paper, wool or hay, arranged to fit snugly round the pot. The box could even be cardboard. The addition of a foil lining helps to reflect the heat back in. Hot boxes will keep the solar-cooked food warm until evening, if it is going to be eaten after dark.

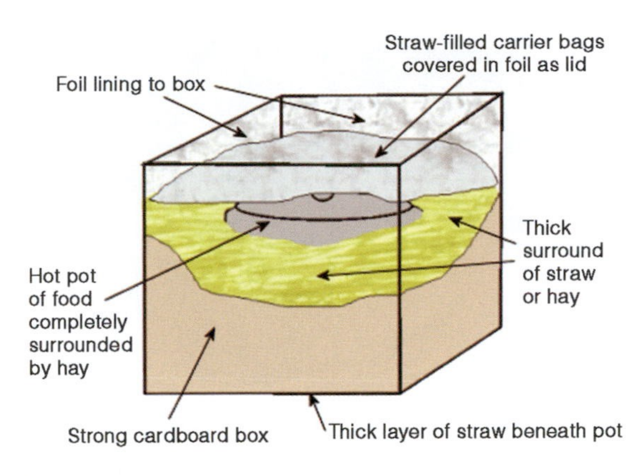

Figure 5.11 A cutaway diagram illustrating the principles of the hot box for cooking.

Products

Various solar cookers are available to buy. For example, one in use in the Himalayas is the SolSource, a lightweight, foldable parabolic shell comprised of several triangular yak-wool canvas panels stretched across a curved bamboo frame and lined with aluminized polyester film (Mylar®).

In India, a solar cooker has been developed by a group called Promoters and Researchers in Non-Conventional Energy (PRINCE). It is a 'square' concentrator made with a series of anodized aluminum strips attached to curved metal bars within a square outer frame measuring 1.2 × 1.2m (3.9 × 3.9ft), which reflects sunlight onto a cooking pot. It is available in bulk from Aadhunik Global Energy.

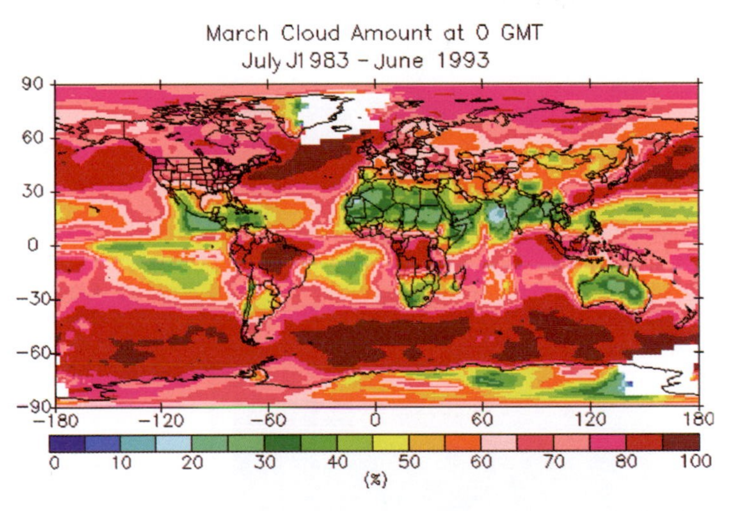

Figure 5.12 NASA has an interactive database that's freely available where you can choose a month and a location in the world and see how much cloud cover there is on average. See Chapter 10 Resources for more information.

Source: NASA

The CooKit solar cooker is another model (Figure 5.13). It is made of cardboard and foil. A heat-resistant bag (or similar transparent cover) surrounds the pot, acting like a greenhouse by allowing sunlight to hit the pot and preventing heat from escaping. It weighs half a kilogram and folds into the size of a big book for easy transport. They were developed by Solar Cookers International, based on a design by French scientist Roger Bernard. With a few hours of sunshine, the CooKit can feed five to six people at gentle temperatures, cooking food and preserving nutrients without burning or drying out. Larger families need two or more cookers. They can be purchased affordably or built using the plans at http://solarcooking.wikia.com/wiki/CooKit.

There also exist several medium- to large-scale parabolic dish and trough-based systems, which heat steam in pipes. The steam is then transferred to a kitchen for use in cooking, or condensed for washing.

Figure 5.13 A CooKit panel solar cooker.

Figure 5.14 This 'Sunny Cooker' can be made yourself.

Source: The instructions are available at http://sunnycooker.webs.com

Figure 5.15 A large parabolic cooker being used in Bangladesh for a family.

Source: Creative Commons

Figure 5.16 A parabolic deep focus cooker in Africa. These pictures are from http://friends.ccathsu.com/bart/solarcooking/parabolic/parabolic_solar_cooker_pg_3_html.htm, where there is a handy applet that allows you to figure out the focal point of any parabola and details of how to design a dish yourself.

Source: Creative Commons

Figure 5.17 Making a solar cooker from a satellite dish. The picture shows a student riveting an aluminium sheet to a satellite dish to make a solar cooker. The holder for the pot was made using part of a bicycle wheel. There is also a counterpoint weight system at the back of the dish made out of old lifting weights that makes it easy to track and position the dish so it points at the sun.

Figure 5.18 Possibly the world's largest solar canteen in Mount Abu, Rajasthan. The cooking system is a joint effort of Brahma Kumaris India and Solare-Bruecke, Germany, The solar reflectors produce steam that's used to cook vegetables and rice for up to 18,000 people. The steam can reach temperatures of 650°C (1202°F) at the focal point of the reflector, hot enough to cook food in massive industrial pots of 200–400l (44–88 gallons). On days of peak solar radiation, the system can allegedly cope with 38,500 meals per day.

Source: From http://tinyurl.com/y4zbeal

Figure 5.19 A trough design, used to make a grill.

Source: Creative Commons

6 Photovoltaics

Electricity is the most versatile form of power that we know of, and if we could capture and turn into electricity just a fraction of the energy we receive from the sun, we would have more than we need. There are several ways of turning sunshine into electricity, some of which use the sun's light and some of which use its heat. The principal technologies are:

- photovoltaics (using its light);
- concentrating solar power (CSP) (using its heat);
- solar towers (using its heat);
- thermoelectrics (using its heat).

Photovoltaics

The word 'photovoltaic' (commonly abbreviated to PV) comes from a Greek word 'photos' meaning 'light' and 'volt' – the SI unit of electromotive force, named after its discoverer, Alessandro Volta. PV solar modules generate electricity from the sun's light. They have no moving parts, require little maintenance, and they produce electricity from a free fuel without polluting the environment.

We will look at the history, the basic principles and then at grid-connected solar electric systems (also known as 'utility interactive' and 'grid-tied') and some other applications and new developments. Stand-alone systems have been given a chapter of their own.

A brief history of PV power

The PV effect was first discovered in 1839 by French physicist Edmond Becquerel. He found that when you shine a light upon certain metals, a stream of particles (later found to be electrons) is emitted from that metal. However, it was only in 1883 that Charles Fritts built the first solar cell, by coating selenium with a thin layer of gold to form the junctions. It was less than 1 per cent efficient.

Experimentation showed that the number of electrons emitted by the metal depended on the intensity of the light beam applied on the metal; the more intense the beam, the higher the number of electrons emitted, but no electrons are emitted until the light reaches a threshold frequency, no matter how intense the light is. What they could not fathom at the time was why the emitted electrons move with greater speed if the light has a higher frequency. The world had to wait for the answer until 1905, when Albert Einstein described the quantum nature of

Figure 6.1 Alessandro Volta, the late 18th-century pioneer in the study of electricity.

Source: Print from the Edgar Fahs Smith collection

Figure 6.2 Alexandre-Edmond Becquerel (1820–1891). Lithograph by Pierre Petit (1832–1885), printed by Charles Jérémie Fuhr.

Source: Wikimedia Commons

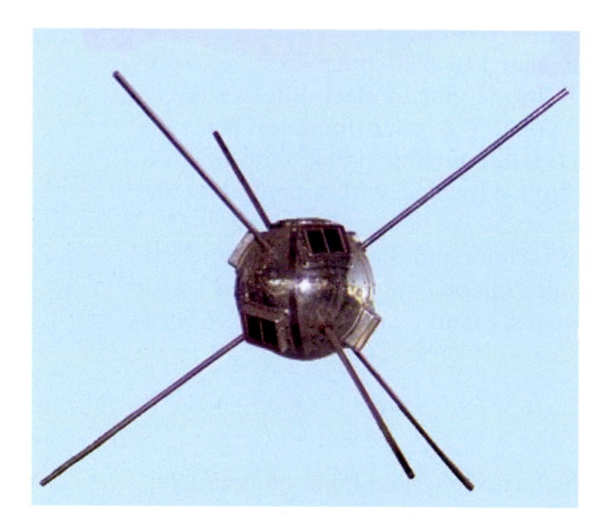

Figure 6.3 *Vanguard I*, the world's fourth spacecraft, and the first to be powered by photovoltaic cells.

Source: Wikimedia Commons

light and its behaviour as a wave as well as a stream of particles (later called photons), for which he won the Nobel Prize in Physics in 1921.

Russell Ohl patented the modern junction silicon semiconductor solar cell in 1946. In the 1950s, Bell Laboratories in New Jersey, US, developed the principle, while researching semiconductors and the effect of light upon them, obtaining efficiencies of 6 per cent. The first use of these novel inventions was to power instruments on board the second US spacecraft, *Vanguard I* (Figure 6.3), in 1958, and thenceforth it was the space race that financed their rapid development.

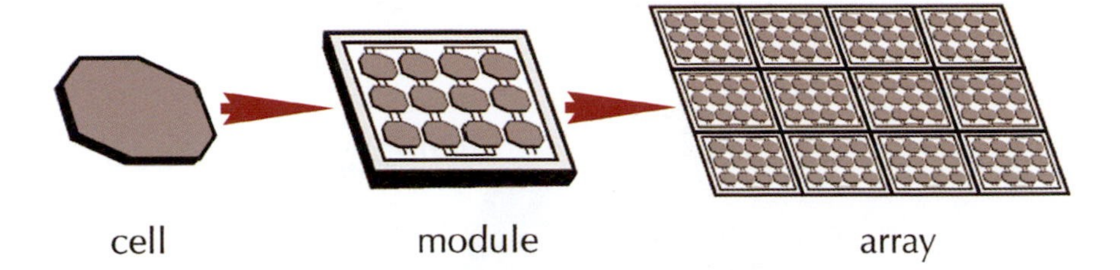

Figure 6.4 Silicon PV modules are arrays of cells connected together and come with electrical outputs varying from a few watts to over 100W of DC electricity. The modules can be connected together into PV arrays for powering a wide variety of electrical equipment.

How a solar cell works

The modules are composed of an array of cells, which are made of semi-conducting metals that have been treated and assembled in a manner similar to the technology underpinning transistors and computer 'chips'. The most common type of solar cell is silicon crystalline, but there are also several other types.

Silicon cells

A standard silicon cell is composed of two layers of silicon, one of the world's most abundant elements. Atoms of silicon have the property that when a photon – a quantum particle of light – hits them, its energy is transmitted to an electron in the outer ring of the atom, knocking it into a higher energy band. The amount of energy required to achieve this is determined by the 'band gap' of the material, which itself affects what portion of the solar spectrum a PV cell absorbs. Ordinarily the electron would soon fall back, as its negative charge is attracted to the positive charge of the atom's nucleus. So, the two-layered structure of the cell is designed instead to capture that electron with minimum energy loss and make it flow in a circuit.

To achieve this, the upper layer of silicon is 'dosed' or injected with atoms of phosphorous, and the lower layer is 'dosed' with boron. These give the layers respectively a negative and a positive potential. The electron that was knocked off the upper layer is now attracted to the lower layer, leaving behind a hole. If the cells are connected in a circuit, this electron, and all the millions of others in a similar state, will flow around it, producing a current. Voltage is created by a reverse electric field around the junction between the layers – which is known as a p–n junction. Returning electrons will fill the original holes – and the process is ready to start again.

Figure 6.5 A monocrystalline solar cell.

Source: Wikimedia Commons

Other metals and compounds besides silicon can be manufactured to have this same property, and modules composed of them are both available and in development. More on this later. Silicon cells are the most widely available at present.

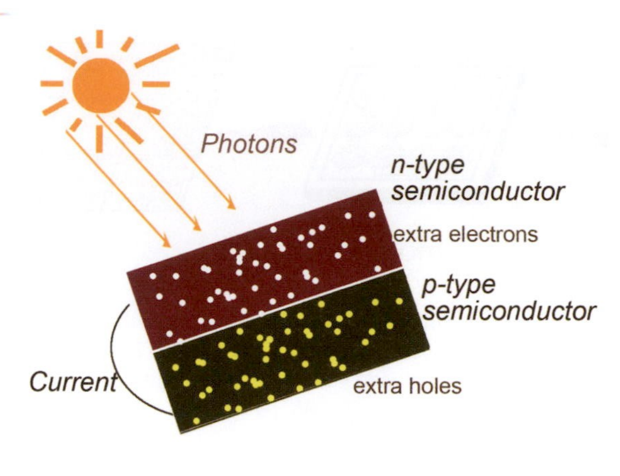

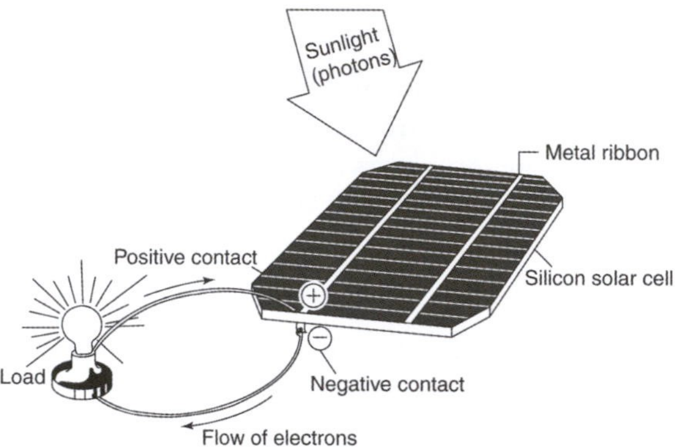

Figure 6.6 When photons strike a silicon PV cell on the negative side of the junction, they dislodge electrons of the same energy level, which move into the positive side of the junction. This creates a voltage difference between the front and rear sides of the cells. If they are connected in a circuit, then a current will flow.

The effectiveness of the process rests upon matching the exact frequency of the light hitting the cells, to their composition. As we saw in the introduction, sunlight is made up of a spectrum of frequencies. The efficiency of a cell is therefore partly dependent on the range of frequencies it can respond to. Photons with insufficient energy will not excite the electrons to jump the band gap. (The higher the frequency, the more energy.) Those photons with more energy than that required will lose this as heat, causing the cell to heat up, reducing its efficiency. In addition, the greater the range of frequencies whose energy can be captured, the greater the cell's efficiency and the more power that will be generated. Extending this range is another significant area of research and development that is reducing the cost of PV. Similarly, since average light quality varies with climate and location, the more the module's characteristics can be matched with the average qualities of the light falling upon it, the more efficient it will be.

Types of silicon cell

Silicon cells can be of three types: monocrystalline, polycrystalline and amorphous. The names refer to the arrangement of the silicon crystals, which is determined by the manufacturing process.

Monocrystalline cells are grown from high quality pure silicon. They are efficient (up to 24 per cent, under laboratory conditions) but more expensive. Polycrystalline cells are made from silicon that is melted and cast, and contains

Figure 6.7 Stages in the manufacture of a silicon solar cell. From left to right: sand containing the silicon, a lump of silicon, a cube of silicon and a finished module.

Source: Frank Jackson

many crystals but are slightly less costly to make and slightly less efficient but are the most common type, representing about 85 per cent of the market.

Thin-film modules

So-called 'thin-film' modules are a more recent development that are gaining in popularity. They work in a similar way to silicon cells but are constructed differently, by depositing extremely thin layers of photosensitive materials onto a low-cost backing, such as glass, stainless steel or plastic, making them potentially flexible and useful in consumer applications including mobile phones. They are more tolerant of shade or higher temperatures than other sorts of cells.

Thin film modules are less costly to manufacture than the more material intensive crystalline technology, but they are less efficient – 5–13 per cent (requiring 11–13m^2 (118–140ft^2) for 1kWp). Therefore, to obtain the same output of power, double the surface area of thin film modules would be needed compared to crystalline modules, which is not always available. Thin film modules are produced in sheets sized for specified electrical outputs. Three types of thin film modules (depending on the active material used) are commercially available at the moment:

- amorphous silicon (a-Si);
- cadmium telluride (CdTe);
- copper indium selenide (CIS) / Gallium diselenide/disulphide (CIGS).

Amorphous silicon cells have no crystals and are less efficient, but they are cheaper as they use fewer raw materials. They are used in pocket calculators and watches. A film of the amorphous silicon is deposited as a gas on a surface such as glass, aluminium or plastic. Thin-film, commercially-produced CdTe and CIGS modules achieve around 10 per cent efficiency. Tellurium (used in CdTe) is a rare substance, and cadmium and gallium do not occur naturally in sufficient quantities to support a mass roll-out of cells utilizing them.

Multi-junction cells (a-Si/m-Si) include not one but several p–n junctions connected in series. The purpose of this is to capture more frequencies of sunlight. Different materials are used at each junction so the junctions have different band gaps and can therefore receive photons with different wavelengths. Efficiencies are currently in the region of 9 per cent outside of the laboratories, but could go much higher in the future.

Power output

The amount of power produced by a solar cell depends on how much light is hitting it. It will perform at its best when pointed directly at the overhead sun on a bright, clear day. This maximum power output is called its 'peak power' (watts-peak or Wp).

Table 6.1 Surface area needed for different module types to generate 1kWp under standard test conditions (STC).

Material	Surface area needed for 1kWp
Monocrystalline silicon	5–7m² (54–75ft²)
Polycrystalline silicon	6–8m² (65–86ft²)
Amorphous silicon	11–17m² (119–183ft²)
CdTe thin film	9–11m² (97–119ft²)
CIS/CIGS thin film	8–10m² (86–108ft²)

Source: IEA

Box 6.1 The making of polycrystalline solar modules

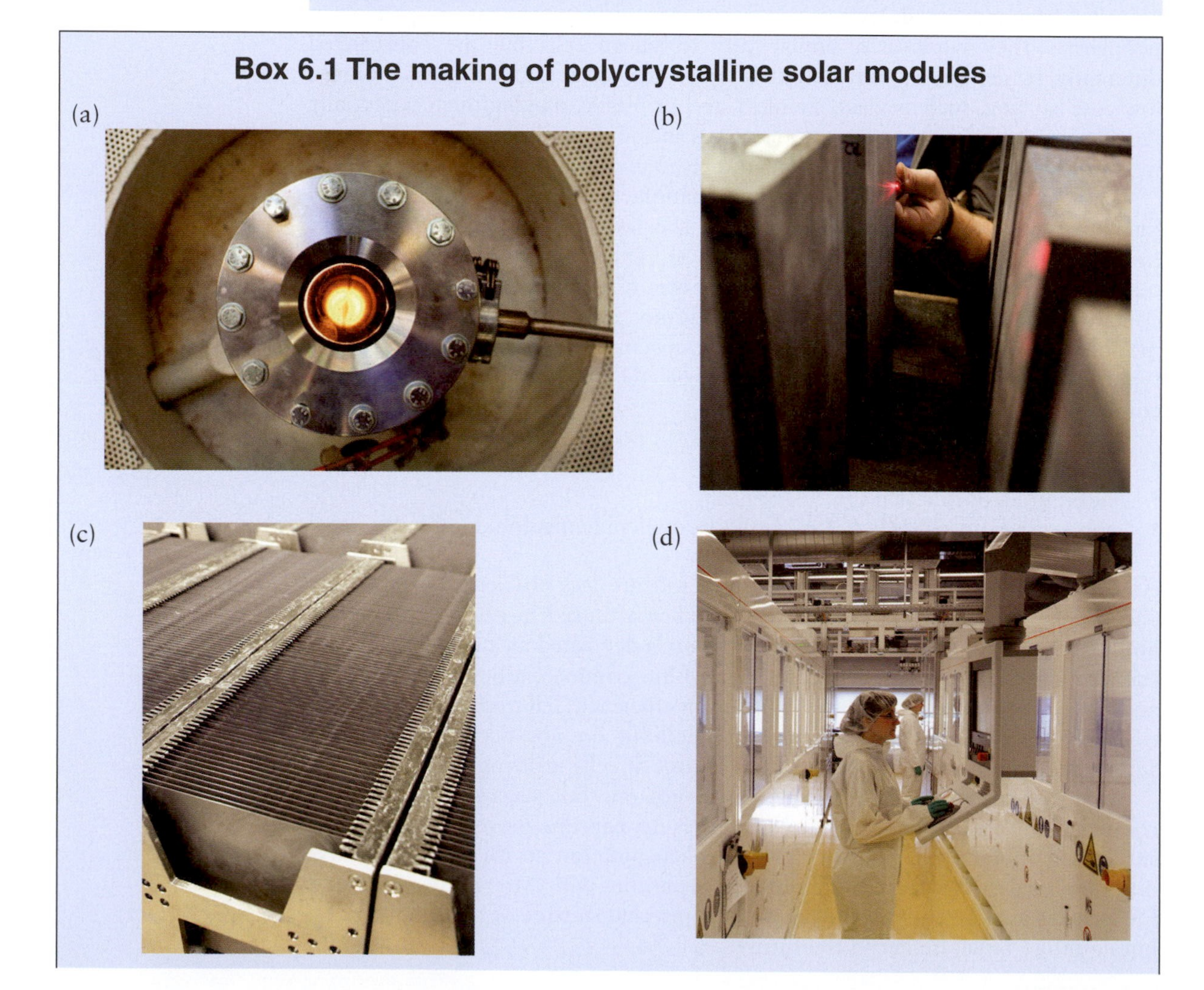

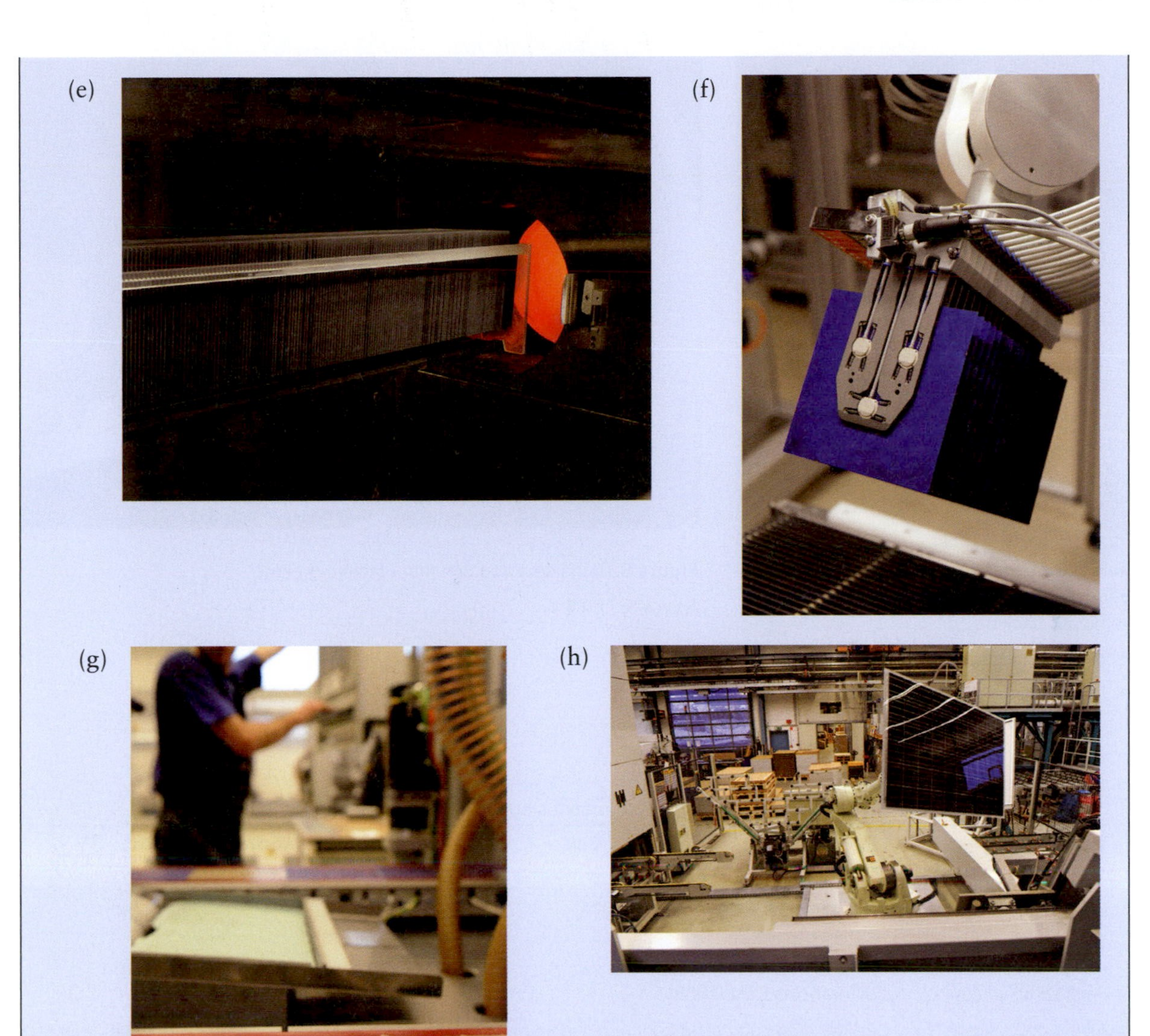

Figure 6.8 The making of polycrystalline solar modules.

Source: www.solarworld.de/4341.html

(a) High-purity silicon is melted in crucibles at more than 1400°C (2552°F). (b) As it cools, the crystals melt in a vertical direction, starting at the bottom of the crucible and forming a polycrystalline block. (c–d) The blocks are then sawn into rectangular wafers and cleaned with chemicals. (e) The p–n junction is created using phosphorous diffusion; this makes the silicon conductive to electrons. (f) A blue anti-reflection layer is added to reduce optical losses and make for an electrical passivation of the surface. (g) The solar cells are then soldered into rows and columns by connecting the front of one cell to the rear of the next cell, in this case to create a module of 60 cells. (h) The module is given a covering of solar glass, films, and a backing plate with a socket on the rear of the laminate for the electrical connection. The laminates are then pressed into an aluminium frame, and the finished modules are tested for quality.

Figure 6.9 A polycrystalline wafer on a laminate backing.

Source: Wikimedia Commons

Figure 6.10 Inside a thin-film manufacturing plant.

Source: © First Solar

Table 6.2 Efficiency and development stages of different PV cell types.

Type of cell	Construction	Cell efficiency (%)	Module efficiency (%)	Current stage of development
Monocrystalline silicon	uniform crystalline structure – single crystal	24	14–20	industrial production
Polycrystalline (multi-crystalline) silicon	multi-crystalline structure – different crystals visible	18	13–15	industrial production
Amorphous silicon	atoms irregularly arranged thin-film technology	11–12	5–9	industrial production
Cadmium-telluride	thin-film technology	17	9–11	industrial production
Copper-indium-selenide	thin film, various deep-position methods	18	10–12	industrial production
Gallium-arsenide	crystalline cells	25	—	produced exclusively for special applications (e.g. spacecraft)
Gallium-arsenide, gallium-antimony and others	tandem (multi-junction) cells, different layers sensitive-different light wavelengths	25–31	—	research and development stage
Organic solar cells	electric-chemical principle based	5–8	—	research and development stage

Note: NB: From the practical point of view of evaluating systems, the module efficiency should be used.

Modules are defined by their peak power output under standard test conditions (STC). These are when the light shining on them is 1kW per square metre, the temperature of the module is 25°C (77°F) and the air mass (the thickness of the atmosphere through which the light has to travel) is 1.5. This standard allows models to be compared. A module that is 100cm^2, or 10 × 10cm (3.94 × 3.94"), which is 12 per cent efficient, will produce 1.2W under STC. Therefore it would be rated at 1.2Wp. If it was three times as big it would produce three times as much power: 3.6Wp. If ten of them were connected together they would produce ten times as much power (12Wp), and so on. However, what we're really interested in is the output of the whole system, which will be less, and we'll examine this later.

Applications of PV

Grid-connected systems

The most common use of PV is to generate power to be fed into the electrical grid as well as being consumed locally. These systems are called grid-connected, utility-interactive, or grid-tied PV systems and come in all sizes, from the domestic scale right up to large power stations. We will look at these in turn, from the smaller to the larger.

Domestic systems

Grid-connected domestic systems use the grid as a backup when no power, or insufficient power, is being generated by the solar modules. At times when there is superfluous power, it is exported into the grid, making the electricity meter run backwards. In this type of system, a synchronous – or grid-commutated – inverter, converts the DC power from the PV arrays into AC power at a voltage and frequency that can be used by domestic appliances and is acceptable to the requirements of the local network.

Systems are usually sized relative to the budget of the homeowner or the amount of roof space available. The amount of electricity generated is usually

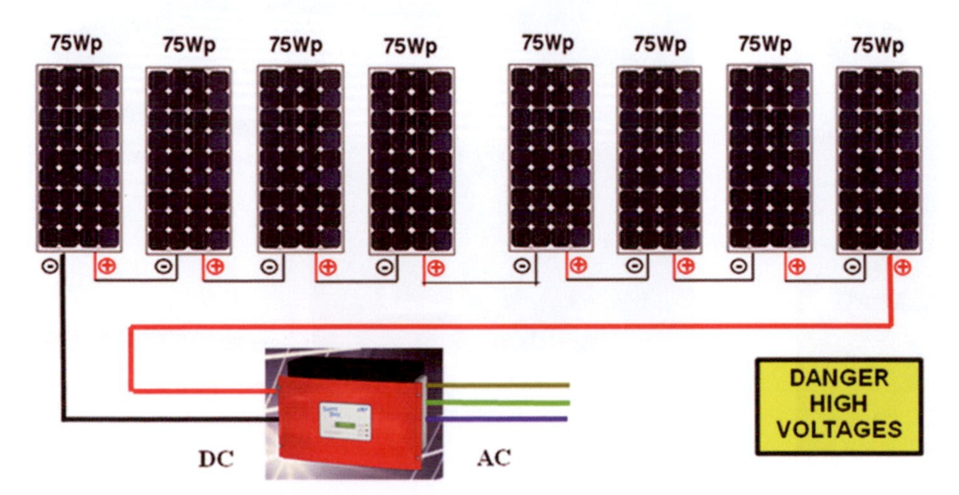

Figure 6.11 Schematic diagram of a system with an array of modules feeding into an inverter.

Figure 6.12 Two inverters installed in a roof space below some modules. The type, size and number of inverters will depend on the PV array configuration, local electrical codes and other factors.

Source: © Chris Twinn

only a proportion of the average yearly total consumed. Homeowners installing PV typically become more conscious of the electricity they are using. The average home might therefore use around 1000–1500kWh per person per year for its appliances and lighting. The commonest modules at this scale are monocrystalline and polycrystalline silicon. Although the former are more efficient, there is no difference in output between the two systems in the real world – a 1kWp polycrystalline array simply has a slightly greater surface area than a 1kWp monocrystalline one.

(a)

(b)

Figure 6.13 Building-integrated solar tiles, which are used instead of normal tiles, on a house in the UK.

Source: © Chris Twinn

Figure 6.14 PV panels and a green roof on a residential boat in Berlin, Germany.

Source: © Frank Jackson

Box 6.2 A passive and active solar house in Wisconsin, US

This is the same house that is described in Chapter 4 on solar space heating, owned by Jo and Alan Stankevitz in Minnesota. Electricity is supplied from the utility company and a PV grid-intertie (grid-connected) system. When generating, this supplies the house with over 3kW/hr of power. Any surplus is sold back to the local network. During times when the sun isn't shining, power is supplied by the local utility. It has been in operation since June 2004 and provides almost enough power on an annual basis to meet the annual electrical demand. It uses twenty-four 58W modules and four 100W modules to total 4.2kWp, fed through two inverters. Some monitored output figures for winter/spring 2009 are: January (346kWh), February (417kWh), March (445kWh) and April (496kWh).

Further information can be found at http://daycreek.com.

How much power can you expect? In a latitude like England's, perhaps 600–800Wh per year per 1kWp rated output. In Germany, the figure starts around 850Wh per 1kWp per year in the north of the country, rising to 1kWh per 1kWp per year in the south. In Southern California, 1800kWh per year is obtained and in North Africa, Australia and India (which receives on average 300 sunny days a year and 4–7kWh/m^2 per day), output may start at 1600kWh and reach 2000kWh. A recent monitoring trial in Oxford, England, found that even a north-facing roof generated 400Wh/kWp – not that this is at all recommended.

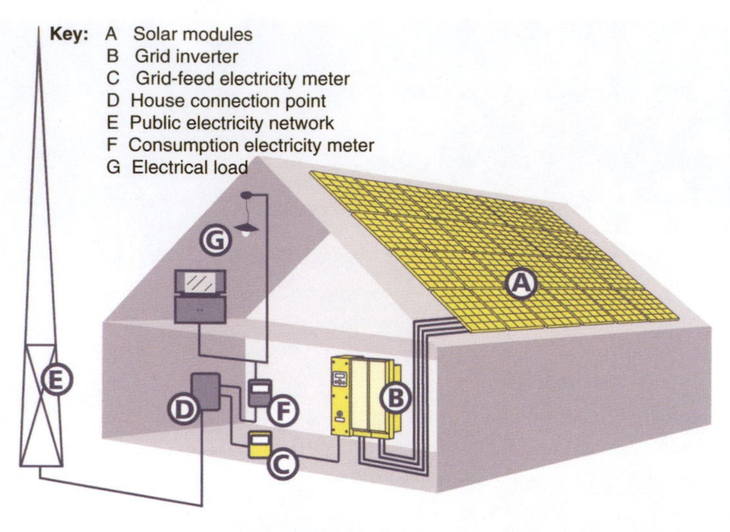

Figure 6.15a The components of a grid-connected system. The modules feed power to the home's appliances through the grid inverter, which produces grid-quality AC electricity. Any surplus not used is sent to the local electricity network through the grid-feed electricity meter. When the home's demand is greater than can be satisfied by the PV array, such as at night time, power is drawn instead from the local network through the consumption electricity meter. The utility company would bill the home for the balance of the amount of power used and supplied, according to the tarrifs agreed for the electricity bought and sold.

Source: Steca

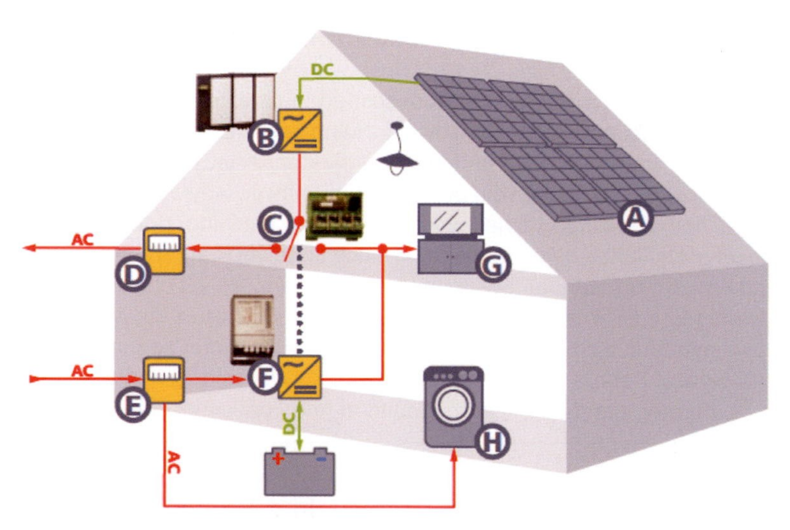

Key:

A Solar modules
B Grid inverter
C Relay module
D Grid-feed electricity meter
E Consumption electricity meter
F Sine-wave inverter
G Supplied loads
H Unsupplied loads

Figure 6.15b The components of a grid-connected system, which will also continue to work on a back-up battery and off the modules in the event of the grid failing and a black-out occurring. This would protect vital services, in, say, a hospital or telecommunications scenario; the relay would automatically switch AC loads, that need to run when the grid is down, to a battery-connected inverter.

Source: Steca

When planning a system, it is advisable to get estimates and proposals from a few installers. PV modules should last 30 years or more, but their power output will decrease over time; the degree of reduction depends on the type of cell. Monocrystalline and polycrystalline silicon modules should come with a guarantee of giving at least 80 per cent of their maximum output after 25 years.

System design

A typical system includes, besides the modules, a grid-inverter, which is a different type of inverter to that used in stand-alone (non-grid-connected) systems, controllers, disconnects, meters and fuses. A system that is also required to supply power when the grid is down would additionally include a battery-connected inverter and relay controller, as shown in Figure 6.15.

It is vital to consider that the annual energy that can be produced by a given array of modules depends upon several factors, particularly the type of module and the location. The yield of a system can be calculated with reference to the annual energy output per peak kilowatt of rated power, which will be supplied by the manufacturer of the module.

The insolation data for the location will give the average amount of energy available per square metre for that location – in kWh/m^2 – for the whole year. A simple calculation will then obtain the estimated output per year. The approximate energy produced per PV module in a year is given by the following formula:

Module rated output (peak watts) × peak sunshine hours (per year) × 0.75 (*performance ratio*) = energy generated (kilowatt-hours/year)

The *performance ratio* is to cover higher module operating temperatures, system losses and other factors – and in a well installed grid-connected system it is usually between 0.7 and 0.8.

For example, 1500W × 900 peak hours per year × 0.75 = 1,012,5005 watt-hours (Wh) or 1102 kilowatt-hours (kWh) (approximately)

The annual total amount of solar radiation available at the site and the peak watt rating are not the only factors that need to be taken into account in system sizing and determining the true final output. Other factors include:

- the orientation (azimuth) (–90° is east, 0° is equator-facing and 90° is west);
- the tilt angle of the modules from the horizontal plane, for a fixed (non-tracking) mounting;
- the energy conversion efficiency of the modules;
- the extent to which their efficiency is affected by temperature;
- factors to do with the clarity of the atmosphere and the path of the sun;
- whether there are any obstructions which might cause shade on the modules at any time of the day and year;
- system efficiencies, such as those of the chosen inverter and the wiring;
- the type of mounting structure – fixed or tracking (tracking increases output).

PVGIS estimates of solar electricity generation

Location: 52°31'24"N, 13°24'41"E, Elevation: 40m

Nominal power of the PV system: 1.0kW (crystalline silicon)
Estimated losses due to temperature: 8.1% (using local ambient temperature)
Estimated loss due to angular reflectance effects: 3.0%
Other losses (cables, inverter etc.): 14.0%
Combined PV system losses: 23.3%

	Fixed system: inclination=36° orientation=0° (Optimum at given orientation)			
Month	Ed	Em	Hd	Hm
Jan	0.88	27.4	1.04	32.4
Feb	1.71	47.9	2.07	57.9
Mar	2.28	70.7	2.82	87.4
Apr	3.19	95.8	4.14	124
May	3.87	120	5.17	160
Jun	3.44	103	4.69	141
Jul	3.65	113	5.00	155
Aug	3.43	106	4.68	145
Sep	2.64	79.3	3.47	104
Oct	1.92	59.6	2.43	75.2
Nov	1.07	32.0	1.29	38.8
Dec	0.63	19.6	0.75	23.2
Year	2.40	72.9	3.13	95.3
Total for year		875		1140

Ed: Average daily electricity production from the given system (kWh)
Em: Average monthly electricity production from the given system (kWh)
Hd: Average daily sum of global irradiation per square meter received by the modules of the given system (kWh/m2)
Hm: Average sum of global irradiation per square meter received by the modules of the given system (kWh/m2)

Performance of grid-connected PV

PVGIS estimates of solar electricity generation

Location: 52°44'7"N, 13°21'33"E, Elevation: 61m

Nominal power of the PV system: 1.0kW (crystalline silicon)
Estimated losses due to temperature: 12.7% (using local ambient temperature)
Estimated loss due to angular reflectance effects: 2.9%
Other losses (cables, inverter etc.): 14.0%
Combined PV system losses: 27.1%

	Fixed system: inclination=45° orientation=25°			
Month	Ed	Em	Hd	Hm
Jan	0.82	25.4	1.01	31.4
Feb	1.60	44.9	2.05	57.3
Mar	2.12	65.8	2.76	85.7
Apr	2.96	88.9	4.06	122
May	3.57	111	5.04	156
Jun	3.15	94.6	4.52	136
Jul	3.34	104	4.82	150
Aug	3.16	97.9	4.56	141
Sep	2.48	74.4	3.44	103
Oct	1.80	55.8	2.39	74.1
Nov	1.00	30.1	1.27	38.2
Dec	0.61	18.9	0.75	23.2
Year	2.22	67.6	3.06	93.1
Total for year		811		1120

Ed: Average daily electricity production from the given system (kWh)
Em: Average monthly electricity production from the given system (kWh)
Hd: Average daily sum of global irradiation per square meter received by the modules of the given system (kWh/m2)
Hm: Average sum of global irradiation per square meter received by the modules of the given system (kWh/m2)

Figure 6.16 Sample of data obtainable from the PVGIS software for the estimated power generation in Berlin, Germany, and Armenia for the same specific 1kWp PV array. See Resources (p209) for details of the software.

Source: Generated using PVGIS web-based software

The result may be that 60–70 per cent of the rated power of the module may be finally obtained. Good design and installation are therefore essential.

Figure 6.17 offers a rough guide to the tilt angle required to produce the optimum power output in grid-connected systems depending on latitude.

Temperature and module performance

Temperature affects the performance of a cell, as shown in the graph (Figure 6.18). Beyond 25°C (77°F), each degree Celsius of temperature rise causes a drop of around 0.5–0.6 per cent power output. At 38°C (100°F), a

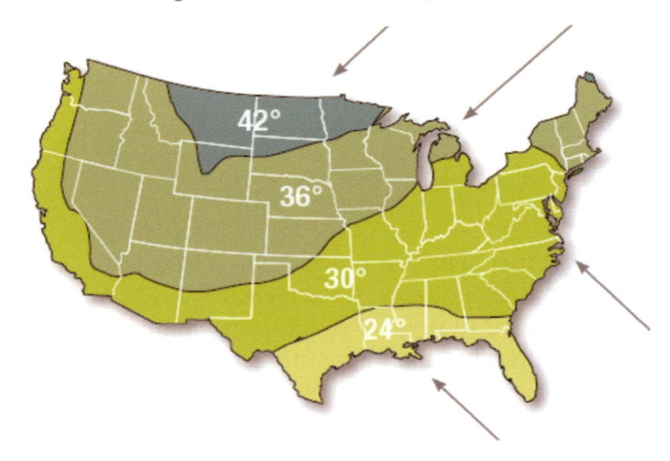

Figure 6.17 Optimum tilt angles for grid-connected PV systems in the US, from US DoE *Best Practice Guide for Solar Thermal & Photovoltaic Systems.* Useful online software for angle selection is PVWatts for North America and PVGIS for Europe (see Resources). It is assumed here that the PV array is facing due south. Angles for off-grid and hybrid systems will be different.

Table 6.3 The effect of orientation and tilt on PV electricity-generating potential.

Orientation	Tilt	Generation (kWh/m²/day)	Difference from optimum (%)
South	35°	3.00	optimum
South	vertical	2.18	17
South	horizontal	2.63	12
East/West	35°	2.46	18
East/West	vertical	1.63	47
East/West	horizontal	2.63	12

Note: The output of a PV array in this location (an apartment block in Manchester, England, with no shading) is reduced by 17% when placed vertically as a building facade, according to these simulation results from the RETscreen software (see Chapter 10 Resources). If they are also east–west facing, then nearly half of the potential power is lost.

Box 6.3 Tips for module positioning

- Position modules where they are never shaded at any time in the year;
- Position them at the correct angle to the sun for your latitude (this is different for grid-connect and off-grid systems);
- Face them towards the equator (usually, these can be exceptions to this);
- Position them where they can be reached easily for cleaning and maintenance;
- Use a mount made of durable, corrosion-free materials;
- High temperatures reduce the efficiency of the modules. Therefore allow appropriate ventilation behind the modules to dissipate heat.

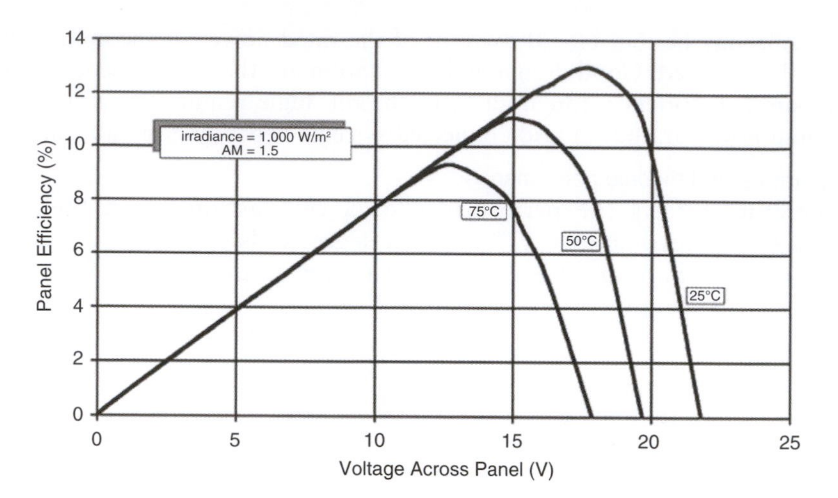

Figure 6.18 How temperature affects the performance of a PV cell. This cell is receiving 1000W/m² in each case (the three curves on the graph), but the higher the temperature, the sooner the voltage drops off and the less power it produces.

crystalline module will produce 6 per cent less power than under STC. The effect is less pronounced for thin-film modules, for which it would be 2 per cent less. Much effort goes into trying to prevent modules becoming too hot in hot weather/climates. One important factor is the movement of air behind the modules; if this is restricted, modules can overheat, so it is important to ensure ventilation behind them. If they are mounted on a rack, the air can move freely behind. If the modules are building-integrated, and are completely built into the structure of the wall or roof of a building, they must be ventilated behind. If the system is in a very windy area, this will reduce the temperature of the modules, which will help to slightly increase their efficiency.

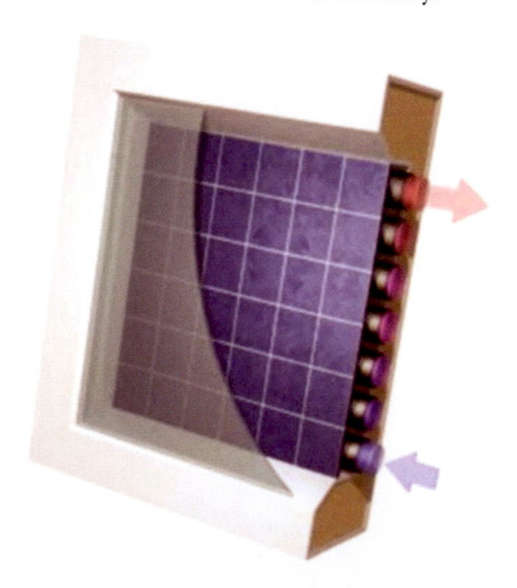

Figure 6.19 A cross-section through a liquid-based flat-plate PV/T module. The PV cells sit above the thermal collector, which helps to cool the cells.

Source: © IEA-SHC

Solar electric-thermal collectors

An attempt has been made to tackle this problem with a new, hybrid solar collector that incorporates both PV power generation and SWH (thermal) – called PV/T. The modules have other potential advantages. The IEA has concluded that PV/T can generate more energy per unit of roof area than the equivalent side-by-side PV modules and solar thermal collectors at a potentially lower production and installation cost. It is claimed that this is because the solar thermal inlet cools the PV component to increase their efficiency by up to 20 per cent, at 45°C (113°F), compared to an identical PV system alone. Various types of commercial products are available, including liquid and air collectors for the thermal component. One solution, for example, is a solar wall with PV cells on top. Another is a panel containing a solar air collector with smaller PV cells driving a ventilator, which sends fresh hot air into the residence. Tracking modules are available too; there is a one-axis tracking, modular, concentrating trough collector. The most common model probably incorporates a conventional tube collector, which can be integrated in a roof.

Box 6.4 A solar wall PV/T system

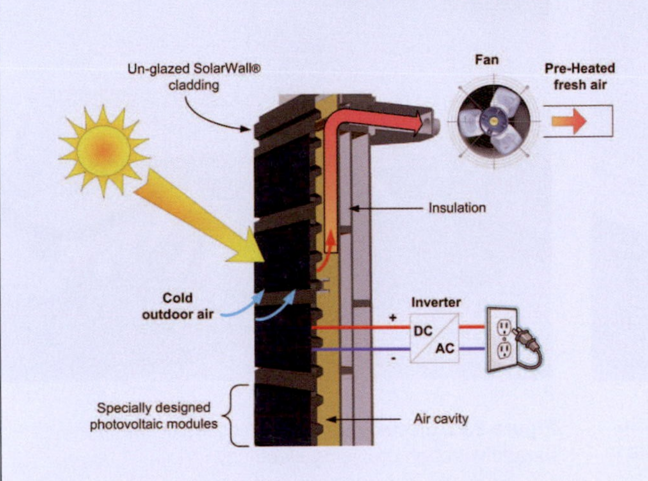

Figure 6.20 The John Molson School of Business (JMSB) is a 24.5kWp building-integrated photovoltaic/thermal (BIPV/T) installation in Montreal, Canada, that uses a high-efficiency distributed air inlet method to recover heat from the PV panels and feed up to about 424,650 litres per minute (15,000ft^3 per minute) of pre-heated air into the building's HVAC system, to offset some of the fresh air heating requirements and cool the panels. The installers say it will yield up to approximately 100MWh of heating per year. The efficiency of the 288m^2 of PV panels can be increased by this cooling, by up to 5% on cold sunny days, compared to an installation without the cooling effect.

Source: NSERC Solar Buildings Research Network (SBRN)

Figure 6.21 The PVTwin PV/T collector; pipes behind the PV module heat water for use, while cooling the module to improve its performance.

Source: © IEA

Figure 6.22 Electric-thermal joint collector with the parabolic trough providing electricity.

Source: © IEA

The IEA's PV/T task group regards it as a very promising technology, with large potential for being applied in a significant proportion of the current solar thermal market, including domestic hot water systems. In the short term, multi-household buildings may be an important market, due to the limited roof area available per household. In the medium and long term, the most promising application seems to be domestic water heating and space heating. This is especially the case with advanced low-energy houses aiming to cover a large part of their energy needs with solar. The combination of a heat pump and PV/T could also be promising. In the longer term, the IEA sees more commercial applications in industry, commerce and agriculture and applications such as solar cooling.

For this to happen, greater standardization is required for performance and reliability, and financing schemes specifically targeted at this hybrid technology. Installation costs need to be reduced together with procedures for PV/T to become an integral part of building design. To this end, the IEA has developed a number of demonstration installations. This is as yet a niche market, however, it is theoretically an attractive concept – although hard to achieve technically – throughout all of the seasons. It would need to be installed by someone competent to meet the electrical and plumbing/heating standards in a given country.

Financial support for solar PV

Usually, different tariffs apply for electricity sold to and bought from the utility company. Some countries have introduced 'feed-in' tariffs (FITs) that pay the

(a)

(b)

(c)

(d)

Figure 6.23 Roof-mounted grid-connected/utility-interactive PV system on the Renewables Academy (RENAC) in Berlin, Germany. The system consists of 3 PV arrays of approximately 1 kWp each. (a) Solon Blue 230/7 200 Wp polycrystalline silicon modules; (b) Sunset TWIN 140 140 Wp multi-junction thin-film silicon modules; (c) Inventux X115 115 Wp micromorph (a-Si/μc-Si) modules; (d) 3 SMA Sunny Boy inverter SMA 1100 inverters, one on each phase. Telecom line for system monitoring goes to the RENAC offices. Three different PV array types were chosen for training purposes. More information is available at www.renac.de.

Source: (a–c) Frank Jackson; (d) Alberto Gallego

supplier well over the going rate for solar electricity that is 'fed into' the national grid. This is a type of subsidy that can act as an incentive to install a renewable energy system, which will therefore pay for itself more quickly. Some governments, such as the UK's, even give this tariff for all electricity generated, regardless of whether it is exported into the system or not. Those countries

Table 6.4 Price summary for solar PV tariffs worldwide (ranked by tariff and years offered; prices at Feb 2010).

Jurisdiction	Application	Years	Tariff €/kWh	1.413 CAD/kWh	1.359 USD/kWh
Italy	Rooftop	—	0.54	0.763	0.734
South Korea	>3kW	15	0.566	0.8	0.77
France	Building Integrated	20	0.42	0.593	0.571
Germany 2009	Rooftop	20	0.43	0.608	0.585
Czech Republic		15	0.487	0.688	0.662
UK	≤10kW new	25	0.4082	0.5479	0.5168
Spain (2007 RD)	<100kW	0	0.34	0.48	0.462
Austria	<5kW	12	0.46	0.65	0.625
Ontario	—	20	0.568	0.8	1.09
Washington State*	Mfg. in state	8	0.846	1.195	1.15
California*	Commercial	5	0.371	0.681	0.504
South Australia*	Residential	5	0.29	0.41	0.394

* Form of net-metering

that have introduced such a system, such as Germany, have seen a vibrant growth in the number of domestic installations with a knock-on effect on jobs in the sector. However, if the tariff is withdrawn suddenly, as happened in Spain in 2009, the industry will contract. FITs are financed either from general tax revenues or from an increase in general average electricity bills. Questions have been raised over the social equity of this policy.

Building-integrated photovoltaics (BIPV) and PV on buildings

Modules can form part of the roof, cladding, overhanging shading or windows of a building, dependent upon the building's location and orientation. It is often most economic to integrate PV with such a building element at the same time as installing or replacing it. The extra cost of the modules and system is partly offset by the cost of the element. For optimum returns, the design of the generation system should be integrated with the whole building design to minimize the total energy consumption of the building. In other words, it should not simply be 'bolted on' afterwards.

Commercial products currently available include: specially designed modules, roofing slates and standing seam metal roofing, and cladding products. Some solar PV companies – for example, in Japan – have gone as far as either buying housing or construction companies, or forging strategic alliances with such companies in order to make it easier to spread the technology. Special attention must be paid to ensure that these products are installed properly to avoid leaks when it rains, do not compromise the thermal envelope of the structure and they carry the necessary fire ratings.

Figure 6.24 Building-integrated solar modules forming the roofs of homes in Freiberg, Germany.

Source: German information portal on renewable energy

Figure 6.25 Transparent thin-film solar modules help with shading to prevent overheating and glare inside this German port building, as well as generate electricity. From the outside, they are difficult to see through but from the inside, they allow an almost unobstructed view out. These modules come as a frameless laminate that can be applied to a building element. They are manufactured using a laser production process, which enables the creation of square recesses to produce the required light transparency, yielding efficiencies of up to 13.8% despite a transmittance of 10%.

Source: BSW-Solar/Sunways

Figure 6.26 The Zero Energy Media wall in Xicui Entertainment Complex, Beijing, China, which has LEDs powered by PV cells embedded in a glass wall, that produce digital art at night time and are powered by solar energy stored from the daytime.

Source: Solarway

Figure 6.27 Semi-transparent PV cells on the roof of the Hauptbahnhof in Berlin.

Source: BSW-Solar/Langrock

Independent monitoring of existing buildings where BIPV or solar cladding retrofit systems have been installed is rare, but what research has been done has shown that they produced less energy than predicted by the designers. There can be several reasons for this: poor siting – imperfect orientation, tilt angle or shading; a lack of control software or appropriate control logic to allow the technologies to work well together; design teams were too optimistic about the ability of the occupants to manage the system and their energy consumption; inappropriate inverter selection; and module–inverter mismatching.

Systems must have on-going and simple monitoring to ensure that problems are quickly identified and resolved. Special care must be taken to make the wiring accessible at the same time as providing a high standard of weatherproofing and airtightness for internal air quality and temperature control. Installation should be carried out by suitably experienced and qualified contractors. A UK government survey of existing installations concluded that: 'even small system sizes of around 1.6kWp can contribute a significant fraction to the building demand. The majority provide 20–80% with an average of 51%'. It also said: 'To maximise the contribution of a PV system to the building load, it is important to combine the PV with other renewable technologies and energy-efficient practices. This may involve training of the building occupants.' It recommends that purchasers of systems obtain a guarantee from suppliers that their PV systems will generate at a certain level of performance.[1] Such guarantees are increasingly forming a part of contracts.

BIPV has a brighter future in some countries than others, depending on the planning system and political will. The Korean Government is hoping to equip 100,000 houses and 70,000 public/commercial buildings with PV systems by 2012; some of the larger projects will qualify for Clean Development Mechanism (CDM) credits, under the global climate change agreements. China is also forging ahead. A law passed in 2006 on energy-efficient construction in the city of Shenzhen means that construction projects which are unable to use solar power will require special permission from the Government to be put on the market. The Shenzhen Construction Bureau now expects 20 per cent of all new buildings to install PV electricity generation systems and for half to have SWH systems. Jiangsu Province installed 50MW of building and rooftop installations in 2010, including 40MW of rooftop projects. At the time of writing there were plans to install a further 200MW, including 180MW of rooftop projects, in 2011.

Cost effectiveness

Temperate zones

Costs are coming down at a significant rate. In Germany, where feed-in tariffs have been in place for longest, the cost in 2010 is now assumed to be around €3250 per kilowatt of installed power. This figure is minus value added tax (VAT), includes insurance and maintenance costs and allows for the annual decrease in performance of the modules. It gives or assumes an annual electricity yield in central Germany of 500–900kWh per kilowatt installed. Further south in the country, better results would be obtained.

Box 6.5 Vista Montana near-zero energy development, California, US

Figure 6.28 Vista Montana near-zero energy development.

Source: © IEA and Consol

This purpose-built estate, constructed in 2003 in Watsonville, California, US, contains 177 detached homes, 80 townhomes and two apartment blocks with 132 flats. The south-facing roofs are covered with monocrystalline silicon PV modules that together generate 1400kWh per installed kilowatt per year. Each dwelling unit will receive 1.2–2.4kW. This is one of many 'zero energy home' developments by Clarum Homes Development Company. The homes contain many energy-efficiency features which, together with the solar modules, reduce energy bills by nearly 90 per cent.

It is important from a policy angle to disentangle the economic performance of the technology from the subsidies and other support mechanisms that governments in different countries use for solar power. This is so that governments can make sensible decisions about the cost-effectiveness of support for different renewable energies and strategies to promote energy efficiency, to obtain the best value for money. If the primary aim is to reduce carbon emissions, then the metric for this would be carbon payback per dollar spent.

For example, in the UK, a study compared a BIPV roof and a notional solar thermal system for a residential block in London. The carbon payback for the solar thermal system was found to be two years and the BIPV system had a carbon payback of six years. Simple economic payback times on current energy prices for both systems were over 50 years. Calculations considering an energy price increase of 10 per cent per year reduce the economic payback time for the PV roof to under 30 years. The study found that the costs of reducing overall carbon dioxide emissions using a BIPV roof are £196/tonne CO_2, for solar thermal individual systems at £65/tonne CO_2 and for community solar thermal at £38/tonne CO_2. The current spot market price for CO_2 is around £15/tonne CO_2.[2] This means that installing a community solar thermal system gives the best return on investment. Overheads are reduced at this scale, and the technology is relatively straightforward.

Having said that, there are many examples of community-scale PV projects. This local scale of projects, rather than a building-level scale, usually offers the best paybacks and carbon savings because of the savings to be made on overall system costs. Smart grid operation, peak load reduction, renewable energy credits and reduced system losses all give further added value to developers. Communities with both commercial and residential loads tend to smooth the overall load profile and optimize the on-site use of PV due to closer commercial load. Rural, high-impedance grids may require additional grid countermeasures. The largest such project is Stad van de Zon (City of the Sun) in the Netherlands, which consists of more than 3500 dwellings and approximately 5MW. The second largest project in terms of generation capacity is Pal Town Josai-no-Mori in Japan (553 houses, 2160kW), and the biggest in terms of number of houses is the Olympic Village, Sydney, Australia (935 houses, 857kW). Japan, with a temperate climate, has one of the highest levels of community-scale PV projects. Many of the installations are on detached houses but some are on apartment blocks.

In California, a residential solar energy system typically costs about $8–10 per watt. Where government incentive programmes exist, together with lower prices obtained through bulk buying, installed costs as low as $3–4 per watt – or

Figure 6.29 A community-scale PV scheme in Cosmo-Town Kiyomino Saizu, Japan, built in 2001.

Source: © IEA

10–12 cents per kilowatt-hour (cents/kWh) – can be achieved. Without such programmes, costs (in an average sunny climate) range from 22 to 40 cents/kWh for very large PV systems.[3]

But, as noted, prices are coming down. 'Photovoltaics and other renewable energy technologies are the only ones to offer a reduction of price rather than an increase in the future,' said a 2008 EU report into PV markets.[4] Prices have declined on average 4 per cent per annum over the past 15 years. Increases in conversion efficiencies and manufacturing economies of scale are the main drivers. As the cost of fossil energy also continues to rise, the payback times for PV are decreasing for more and more applications.

Hot or sun-belt countries

It is not surprising to discover that PV is most effective in areas of the world where the sun shines the most. Days and weeks of uninterrupted sunshine upon modules oriented optimally according to each season will provide the best return on investment. The so-called sun-belt countries between the 35°N and 35°S latitudes are not only the best candidates for solar power but are also experiencing huge growth in electricity demand. Sun-belt countries contain 75 per cent of the world's population, and the fastest growing populations.

One example of such an area is Southern California, where the Californian Power Authority has calculated that there is a great match between supply and peaks of electricity demand for air conditioning in the summer. The authority supports the installation of solar electricity for grid connection in the summer months. Specifically during heatwaves, the sun is unobstructed and even early morning and evening skies are powerfully bright. They calculate that an installed dependable PV capacity of 5000MW reduces the peak load from that day by about 3000MW.

Installations are still focused in parts of the world where there is government financial support, but PV is expected to rapidly become a competitive source of energy in sun-belt areas. 'The levelized costs of energy (LCOE) for large-scale grid-connected PV should decrease by 40 per cent by 2015 and by almost 60 per cent by 2020 if the full growth potential of these countries is realized,' predicts a recent report. The LCOE is the total of all costs and benefits over an installation's lifetime. It is used to determine the price at which that system should be sold. 'As a result, PV will outperform peak energy sources by 2015. PV could even be competitive with some medium to base load technologies in many countries.'[5]

There is much talk of both PV power stations and decentralized supply. The 2010 film about solar energy, *Die 4. Revolution – Energy Autonomy* features both scenarios for the future. Some, like the IEA, favour centralized power stations distributing solar power in sun-belt regions – a more sustainable version of today's model. Others, like the late president of Eurosolar and general chairman of the World Council for Renewable Energy, Dr Hermann Scheer, argue that architects and planners need to think differently and integrate solar power into their buildings and streets. He says in the film: 'It is simply lack of foresight on the part of the architects and clients, who do not see that the negligible additional cost will soon be offset through one's own autonomous free electricity supply. In the same way [that] it is not allowed to dump the garbage onto the street in modern-day civilized cities, it should nowhere be allowed to tolerate avoidable emissions and hence to expose the entire society to the associated consequences.' As David Carroll, the director of Wake Forest's Center for Nanotechnology and Molecular Materials and inventor of a new fibre PV cell, says: 'What if you didn't own your roof? What if the power company did', and paid you for the power it generates?

Areas with low population density – rural countries such as China and much of Africa – can benefit from decentralized power generation. The cost and inefficiencies of transmitting electricity over long distances are avoided when it is generated locally. Scheer advocates this model as an application of PV, especially in developing countries. Promoting it is the purpose of the Eurosolar organization, of which he was president. Partly for this reason, and partly to take advantage of the growing global market, China is scaling up production of solar modules for both domestic use and export. Established solar companies in Japan, Europe and the US are facing growing competition.

Box 6.6 Asan Green Village, Korea

Figure 6.30 Asan Green Village, Korea.

Source: © IEA-SHC and S-Energy Co, Ltd

This estate of 26 semi-detached buildings was constructed in 2005. Each one has 8kW of crystalline silicon PV modules mounted on their inclined, equator-facing roofs. The buildings were deliberately designed to face south. Each building houses four families, allowing for 2kW per family. The system is grid-connected. Asan Green Village was constructed by Habitat for Humanity Korea, an affiliate of Habitat for Humanity International.

Grid parity

The point at which PV power becomes competitive with traditional forms of power generation will vary according to the location and whether it is off-grid or grid-connected. Grid parity is when it becomes cheaper for PV system owners to generate electricity from their own PV array rather than buy electricity from a utility. At what point this occurs depends on the location (level of solar radiation), the installed cost of the entire PV system and the price the customer normally pays the utility for electricity. In some places, depending on these circumstances, it could occur quite soon. General Electric's Chief Engineer has predicted grid parity in sunny parts of the US, including California, by around 2015. In Germany, it has been predicted to happen even sooner. Strictly speaking, grid parity does not involve subsidies – it depends on (i) installed cost per Wp, (ii) level of solar radiation and (iii) price for electricity the customer is paying the utility. Keep it simple. Once grid parity is achieved then subsidies are withdrawn.

General Electric's Chief Engineer has predicted grid parity without subsidies in sunny parts of the US, including California, by around 2015. In Germany, albeit with feed-in tariffs, it is predicted even sooner.

Capacity credit

Another way of looking at the economics of energy supply and where solar power is cost-effective is to compare its capacity credit. This measure of performance as a reliable source of peak energy is otherwise called the 'effective load carrying capability' (ELCC). It represents the technology's ability to provide power to the utility company when it is needed. Several studies using

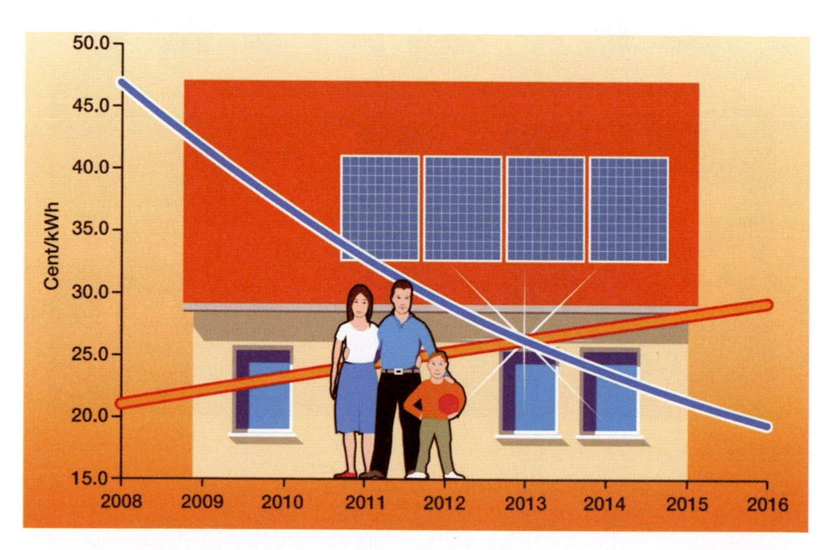

Figure 6.31 Grid parity in Germany is expected in the near future. Grid parity is when it becomes cheaper to produce electricity with your own PV array (represented by the falling line of costs) than to buy it from a utility (represented by the rising line of costs). When exactly this occurs depends on the cost per kWp of the installed PV array, the annual level of solar radiation and the price the customer is paying the utility for electricity.

Source: BSW-Solar, www.solarwirtschaft.de

utility load data show that there is a high capacity credit for solar electricity at various places throughout the US, especially where the conditions feature intense summer heatwaves, high daytime commercial demand for electricity and small electric heating demand.

These conditions now occur in many urban areas around the world. It should be considered essential in these places to also use solar thermal cooling – active solar cooling that uses solar thermal collectors. As described elsewhere, these provide heat energy to drive adsorption or absorption chillers and are far more efficient than using electrical air conditioning and cooling, with or without PV as the electrical source.

The annual average ELCC for PV power plants in California is 64 per cent of the PV system rating. It can go higher than 80 per cent during heatwaves. The Californian Power Authority therefore considers it important to meeting the state's energy reserve needs. Installing PV power generation within the distribution system reduces the amount of power that needs to be transmitted through the grid at these times when compared to locating larger fossil fuel plants at remote locations on the transmission system.

It is important to realize that its success depends very much on the type of load demand and consumption patterns. Demand must be at the time when the peak sunshine occurs, although this can be improved by using tracking systems or a mix of rooftop orientations favouring south and south west, with the south west ones at a higher angle and the southern ones at a lower angle.

PV power stations

In sun-belt areas, it becomes economical to concentrate collection in large power plants. These typically cover large areas of land (or industrial rooftops), often with modules that track the sun as it moves across the sky.

Figure 6.32 The Lieberose solar farm in Brandenburg, Germany, built by Juwi Group and First Solar Inc. on the once-polluted site of a munitions dump. With an output of about 53MW and the size of more than 210 football fields, it was the second largest PV installation in the world on completion in 2009.

Source: © Juwi Group and First Solar

Very precise calculations of the power to be expected are essential if larger systems are being designed. These include more variables than those for domestic-scale systems. For example, they may take into account on a month-by-month basis factors such as:

- the clearness index for the atmosphere;
- whether they track the path of the sun or not;
- the diffuse and direct (beam) radiation amounts;
- the time of sunrise and sunset;
- the angle of these relative to the modules.

There will also be compensation for the reflectivity of the surfaces around the modules and the reflectivity of the surface of the module depending on the angle at which light is striking it. This will allow designers to see whether the amount of electricity generated in a winter month, say, is sufficient, or how many modules are needed to satisfy demand in the cloudiest months.

Tracking systems

In the context of larger installations, it can be more economical to invest in tracking systems, as they can increase their efficiency by up to a third. Tracking allows the PV modules to follow (track) the movement of the sun across the sky, to increase the amount of direct sunlight arriving at the modules. This movement can be made in three different directions:

- Vertical axis: the modules are mounted on a vertical rotating axis, such that the angle to the sun is always as small as possible (this means that it will not rotate at a constant speed during the day).
- Inclined axis: the modules are mounted on an axis that forms an angle with the ground and points in the north–south direction in a plane parallel to the axis of rotation.

Figure 6.33 A two-axis tracking PV system in a large German solar power plant. This module automatically tracks the sun along the east–west and up–down axes.

Source: BSW-Solar/S.A.G.Solarstrom AG

- Two-axis tracker: the system can move the modules in the east–west direction and tilt them at an angle from the ground, so that the modules always point at the sun.

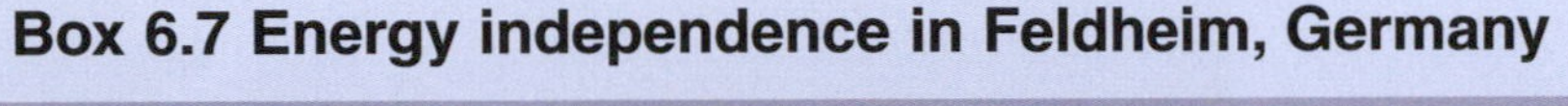

Box 6.7 Energy independence in Feldheim, Germany

Figure 6.34 Energy independence in Feldheim, Germany.

Source: © Frank Jackson

The 145 residents of Feldheim, Germany, are enjoying complete energy independence from corporations with their own system, which combines heat and power from wind, biogas and solar. Besides obtaining 74MW (170 million kWh a year) from wind, and, using a CHP plant running on agricultural residue to generate 4 million kWh of electricity and 4.3 million kWh of heat a year, they also have 2.25MW of PV: 96 arrays, mounted on tracking mounts, are located in a former military training area. Completed in 2010, the exact output is not yet monitored but is expected to be around 3.375 million kWh.

To manage the system, residents formed their own company, Feldheim Energie GmbH & Co KG. Each of them paid €3000 to join, and in return have a guarantee that their prices will not rise beyond 15 per cent less than the commercially available utility rate. In fact, it may well be below that once the €3 million debt – the cost of installing the system – is paid off within 12 years. Such village-scale multi-resource systems are especially appropriate for countries with large rural populations not linked to the national grid.

Where are the existing power stations?

At the time of writing, Spain has the largest PV power plant in the world – 60MW – at the Olmedilla Photovoltaic Park. Germany comes a close second with its 54MW Strasskirchen Solar Park and the Lieberose Photovoltaic Park (53MW). Worldwide, there are 42 operating plants of over 14MW, mostly in these two countries and the majority completed in the last four years. Other pioneering countries include the US, with its DeSoto Next Generation Solar Energy Center in Florida, owned by Florida Power & Light (FPL) and opened by President Barack Obama on 27 October 2009, South Korea and, perhaps surprisingly, Canada (two plants of over 100MW). Over 2GW of further power is in planning or under construction worldwide.

(a)

(b)

Figure 6.35 Olmedilla Photovoltaic Park, Spain, was the largest solar power plant in the world in 2009, occupying 108ha (267 acres). Its 162,000 solar PV panels can generate 60MW of electricity on a cloudless day (i.e. 224.83kWp per acre or 4.45 acres per MW).

Source: (a) Nobesol Levante (b) SMA Solar Technology AG

Concentrating PV

A new generation of PV power stations is emerging, which uses a potentially more efficient process – concentrator solar cells (CPV). CPV installations consist of arrays of concentrating optics that focus and converge sunlight. Modules are classified by how much they concentrate the sunlight, which depends on the optics used. They can range from 4× up to 1500×. The greater the concentration,

Figure 6.36 President Barack Obama opening the DeSoto Next Generation Solar Energy Center, then the largest in the US, at Arcadia, Florida, in October 2009, where he spoke in favour of the "smart grid" as part of the US Recovery Act.

Source: White House website/public domain

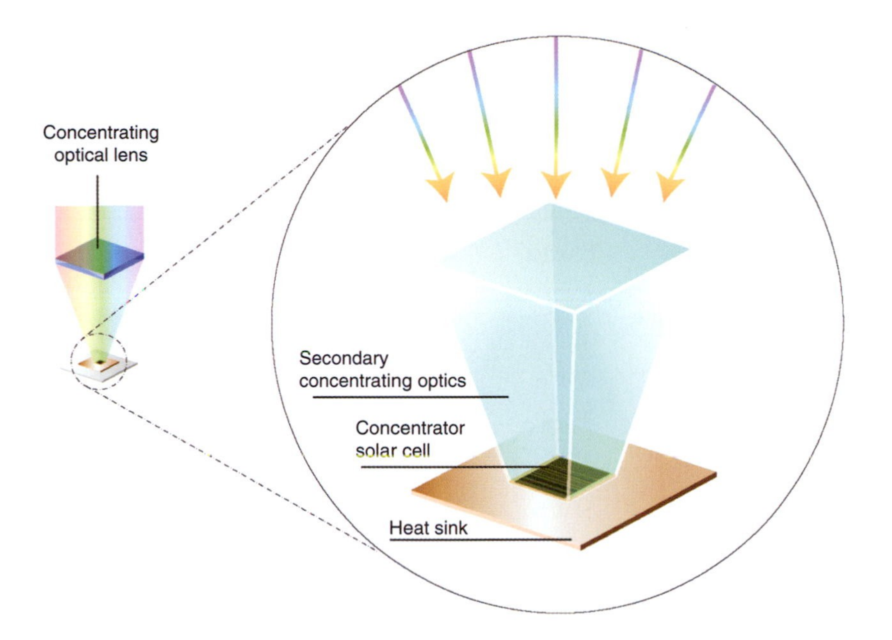

Figure 6.37 Schematic diagram of how CPV works by focusing light onto a small area to target a PV cell. A heat sink helps to lower the high temperature.

Source: Adapted by the author from a CPV brochure

the larger the heat sink that is needed to dissipate the heat generated from the high-intensity light – the equivalent of up to 1500 suns. In the highest concentrations, the target cells are multijunction, high-efficiency, multilayered cells, which make use of a greater spectrum of light, converting up to 40 per cent of light energy into electrical energy. They can be more economical than traditional systems when deployed in the right location: sun-belt, clear sky regions, where normal direct irradiation is greater than 1800kWh/m^2/year. CPV power stations require less land area and fewer cells (and silicon) for the same output.

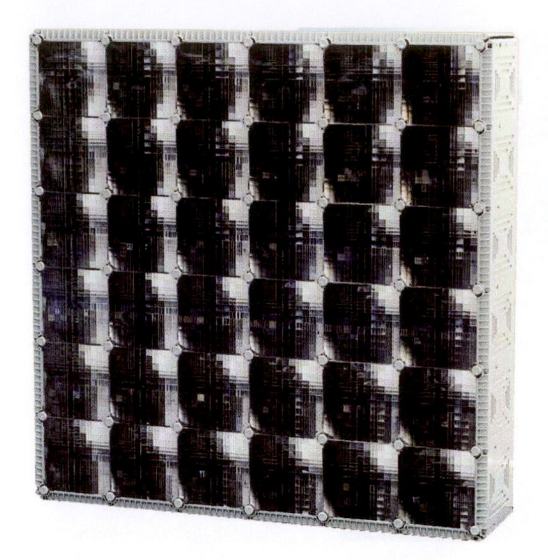

Figure 6.38 A medium concentration CPV module using prismatic lenses to concentrate by a factor of 120 onto 18% efficient monojunction silicon cells.

Though the cost of the cells is greater than that of comparable silicon cells, their cost remains a small fraction of the cost of the overall CPV system, so the economics might still favour the multijunction cells. This depends partly on the module and how powerful the magnification is – at the time of writing, medium-scale concentrators using prismatic lenses, equivalent to around 120 suns, are the most economical. In general, however, solar thermal electricity generating power stations are currently more economical as the technology is more mature. An Australian company, Solar Systems, is producing small power stations using this technology and has a 154MW power station in development using a tower onto which light is reflected. Another company plans to build 34 CPV power stations by 2020 in Australia, each with a peak output of 250MW.

How expensive is PV?

Approximate average retail prices at the time of writing (summer 2010) are as follows:

Module pricing per peak watt	Unit	June 2010	May 2011	% reduction in 11 months
Europe	€/watt	4.13	2.69	65.13%
US	$/watt	4.23	3.07	72.58%
Lowest mono-crystalline module price	$/Wp	2.23	1.8	80.72%
	€/Wp	1.65	1.21	73.33%
Lowest multi-crystalline module price	$/Wp	1.74	1.84	105.75%
	€/Wp	1.29	1.23	95.35%
Lowest thin-film module price	$/Wp	1.76	1.37	77.84%
	€/Wp	1.3	0.92	70.77%

Source: solarbuzz.com

The future of PV

The PV sector is expanding rapidly in many countries. About 5.56GW of PV capacity were installed during 2008 (an increase of about 150 per cent over the previous year) and 7.3GW in 2009, which brought the total installed capacity to 20.7GW. In 2009, 68 per cent of this was installed in the Czech Republic and Germany alone. If Italy, the US, Korea and Japan are included, then over 96 per cent of grid-connected PV installations in 2008 were located in six countries. The biggest off-grid market is India.

According to the European Commission's Joint Research Centre in the Institute for Energy, global production of PV modules increased to about 7.3GW in 2008.

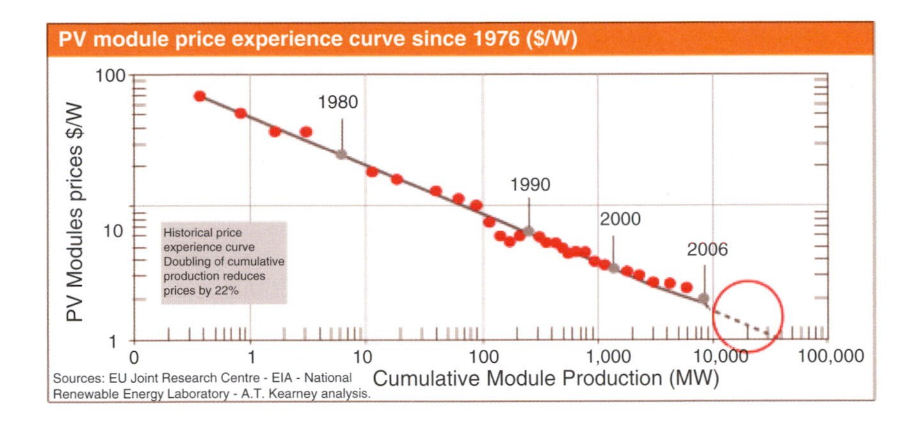

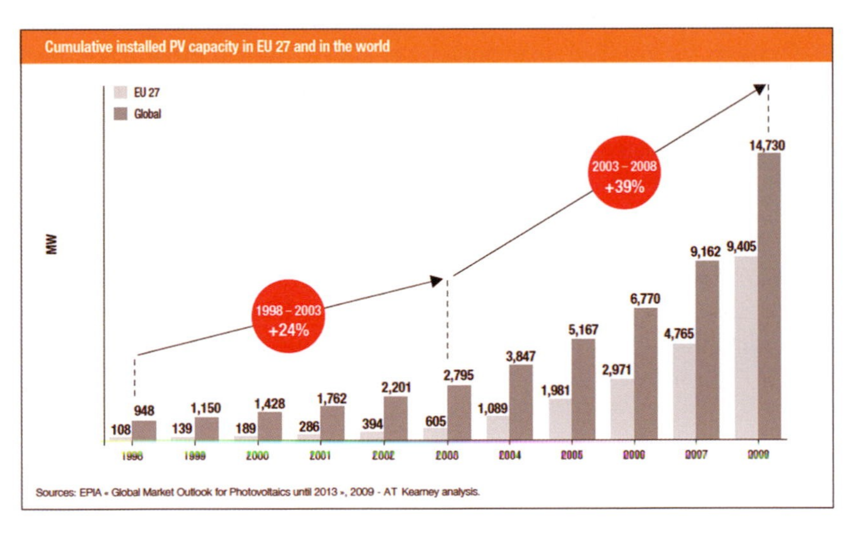

Figure 6.39 Falling prices are matched by the rising numbers of installations. The pattern into the future looks set to continue for solar electricity.

Source: Solar Photovoltaic Electricity: A mainstream power source in Europe by 2020, European Photovoltaic Industry Association, Brussels, 2010, www.setfor2020.eu, accessed July 2010

This is an 80 per cent increase on the previous year, which itself saw a 60 per cent increase on 2006. World solar cell production reached 9.34GW in 2009, up from 6.85GW a year earlier, with thin-film production accounting for 18 per cent of that total; 49 per cent of solar cells were produced in China and Taiwan. It is estimated that by 2050, PV will provide around 11 per cent of global electricity production and reduce 2.3 gigatonnes (Gt) of CO_2 emissions per year.[6] This rate of increase is many times greater than that forecast in a 1996 European plan for the future of solar power in Europe. This envisaged production in 2010 being 1000MW with 2000MW installed, and was at the time considered optimistic. It forecast just a 35 per cent year-on-year growth.[7] This is spectacular growth by any standard and is partly due to government support, notably in Germany.

The sector is investing heavily in research and technological innovation, and generates employment, the majority skilled, high quality jobs. The European Photovoltaic Technology Platform estimates that the PV industry has the potential to create more than 200,000 jobs in the European Union by 2020 and ten times this number worldwide. China is now the leading producer of solar cells with an annual production of about 2.4GW, followed by Europe with 1.9GW, Japan with 1.2GW and Taiwan with 0.8GW. If this trend continues, China might have about 32 per cent of the worldwide production capacity by 2012.

PV products are reliable and last for over 20 years, making them a good investment. At the time of writing, 90 per cent of current production is wafer-based crystalline silicon. Production has recently been held back by a shortage of suitable grade silicon, but this is hopefully a temporary problem. Thin-film cells are increasing in production and, if all expansion plans are realized, could reach '11.9GW, or 30 per cent of the total 39GW in 2010, and 20.4GW in 2012 of a total of 54.3GW' projected.[8]

In Europe, the cumulative installed capacity of PV at the end of 2008 was 9.5GW, more than three times the target it set itself in 2000. In the US in 2008, the cumulative installed capacity was around 1.15GW (768MW grid-connected). The US was the third largest market with 342MW of PV installations in that year, of which 292MW was grid-connected – California, New Jersey and Colorado account for more than three quarters of this.[9] China is aiming for 2GW total installed solar capacity in 2011. In July 2009, a new Chinese energy stimulus plan set 2020 targets for installed solar capacity at 20GW. Similar growth is happening in India (which has a National Solar Mission) and Korea.

To put this in perspective, however, PV generated just 0.35 per cent of Europe's final electricity consumption in 2008 with approximately 1.5 million households in Europe getting electricity produced from solar PV cells. Worldwide PV accounts for less than 0.01 per cent of total primary energy demand. The IEA projects that electricity demand (17.3 trillion kWh in 2008) will rise by 78 per cent to 4.7TWh by 2030. Therefore, even if PV keeps up its current expansion rate, it will only just continue to keep pace with the increase in overall demand. To make a real difference, deployment must increase at an even faster rate, which requires strong, international political will.

New applications

Transport

Solar energy does not generate sufficient on-demand power to substitute by itself for conventional fuels in a vehicle or aeroplane. However, this has not prevented human ingenuity from trying to use solar energy in boats, vehicles and even aeroplanes. There are niche markets that solar-powered transport can fill, and in some cases it is doing so already, particularly in the case of boats.

Solar boats

Some boats, such as pleasure boats, do not need to travel at speed and have a modest load. Solar panels on the canopy can charge batteries in the hold, which can drive an electric engine. Solar can also be used in a hybrid system with wind and diesel, and can be employed just to power the lights and other appliances on board.

Vehicles

For many years, there have been land-speed contests for solar-powered vehicles to drive the technology forwards. The largest of these is the World Solar Challenge, which covers 3021km (1877 miles) through the Australian Outback, from Darwin to Adelaide. The challenge for participants is to maximize the resilience and lightness of the vehicle and improve efficiency, particularly in battery technology.

Figure 6.40 A solar-powered boat in Berlin, Germany. It can carry 16 people.

Source: © Frank Jackson

Figure 6.41 The winner of 2009 Global Green Challenge, 'Tokai Challenger', of the Japan Tokai University Solar Car Team.

Source: Wikimedia Commons

But most of us would not want to commute in a car like any of those competing in the event. So, car makers are prototyping many different solutions, ranging from ultra-lightweight nippers, rather like enhanced buggies, to hybrid solutions, which may incorporate batteries that can be charged overnight from the grid or during the day from PV arrays in the workplace. They are even developing fuel cell-driven models, which may or may not run on hydrogen that has been produced by electrolysis using PV electricity. Among these are General Motor's hybrid concept car, the Chevy Volt, and the Cadillac Provoq Fuel Cell E-flex, which runs on a hybrid electric engine fed by a hydrogen fuel cell. It has

Figure 6.42 Mitsubishi's i-MiEV (Mitsubishi In-wheel motor Electric Vehicle), which uses a series of lithium-ion batteries to power electric motors located in the wheels. Mitsubishi claims they can be charged up directly from solar panels at a workplace. In practice this is unlikely in most of the world – 1 litre of petroleum contains about 11kWh of energy, an amount which would require acres of panels to drive 200 miles a day in such a vehicle, for example. Running cars from solar electricity is unlikely in the foreseeable future as only a small portion of the energy in the global electric grid is generated through solar right now. Grid electricity for electric vehicles is likely to be a mix from many sources.

Source: Mitsubishi

a range of 300 miles for every 3kg fill of hydrogen; it does 0–60mph in 8.5 seconds and has a top speed of 161km/hr (100mph). Its solar roof powers onboard electronics.

Fuel cell, hydrogen and electric hybrid vehicles now also include buses, trucks and trains. For example, Ballard has for many years been running buses in London and the US powered by hydrogen fuel cells.

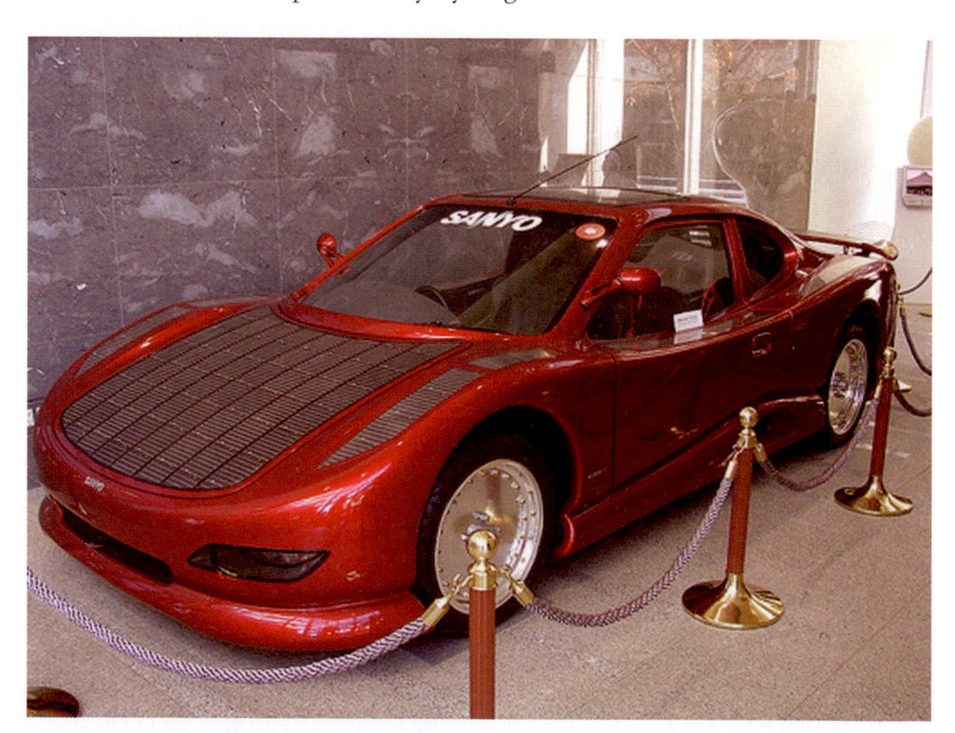

Figure 6.43 Solar panels on the surface of a concept electric car from Sanyo. However, they can only provide sufficient power for the electronics.

Source: Sanyo

Planes

In the field of air transport, major advancements have recently been made. Planes such as the Sunseeker could already achieve what a glider with small motor-driven propellers could do in the daytime. But, in 2010, the first 26-hour non-stop solar-powered flight established that manned planes can fly overnight, and a global circumnavigation is planned. In other words, the weight of the batteries, plane and pilot can be counteracted by the plane's design and self-generated solar power, overnight. The *Solar Impulse* designers are led by André Borschberg and Bertrand Piccard. Its wings are as wide as the span of a Boeing 707, and boast 11,628 PV cells covering 200m² (2200ft²). These charge lithium-ion

Figure 6.44 The Sunseeker solar-powered one-seater, designed by Eric Raymond, has successfully traversed France and Spain.

Source: Eric Raymond

Figure 6.45 *Solar Impulse* making its record-breaking 26-hour flight in July 2010.

Source: © Solar Impulse/ Stephane Gros

Figure 6.46 This unmanned aircraft will be able to stay aloft for weeks and months without landing and has applications in military, surveillance, meteorology and telecommunications. It is likely that the main application for PV in the air will be for military and surveillance purposes.

Source: QinetiQ

batteries weighing 450kg, with a capacity of 200Wh/kg or 90kWh, which power four 7.5kW electric motors. In the daytime hours of the flight, they collect enough energy to power the plane for the entire flight. This is a European project being conducted at the École Polytechnique Fédérale de Lausanne.

The *QinetiQ Zephyr* is a lightweight (30–34kg [66–75lb]) solar-powered plane designed by the UK defence firm, QinetiQ. Constructed with carbon-fibre, its 18m (59ft) wings are covered with solar cells, in this case charging lithium

Figure 6.47 This French solar-powered blimp could be the forerunner of solar-powered freight or leisure flight services.

Source: Projet Sol'R

sulphur batteries during the day to power the aircraft at night. To launch it, five individuals run gently with it into the wind until it wafts out of their hands.

Another pioneering aircraft, named *Nephelios*, is hoped to be a forerunner of tomorrow's niche transport solutions. Semi-flexible PV panels installed on the top side of a 22m (72ft) helium-filled blimp power an electric motor situated behind the steering pod, which turns two twin-bladed propellers enabling it to reach speeds of over 25mph. The French designers, Projet Sol'R, believe that the approach has a future in delivering heavy or bulky freight or leisurely journeys that require high manoeuvrability, quiet operation and zero carbon emissions.

Environmental impact of PV

PV modules, when generating power, produce no pollution or greenhouse gases. However, the manufacturing of them and the system components, their assembly and eventual decommissioning, does have an environmental impact.

Energy payback

The term 'energy payback' refers to the length of time it takes for a PV system to generate the same amount of energy that went into making it. The US Department of Energy (DoE)[10] calculates that the energy payback for polycrystalline modules is four years for systems using recent technology and two years for anticipated technology. However, it depends where they are installed. Sunnier climates like North Africa will pay back their energy sooner than northern Europe. The DoE adds that on average amorphous silicon takes one to two years to generate the energy needed to make it, and three years including the frame and support structure for a roof-mounted, grid-connected system. For off-grid systems, the payback will be much longer.

The DoE refers to findings by Dones and Frischknecht that PV-systems fabrication and fossil-fuel energy production have similar energy payback periods (including costs for mining, transportation, refining and construction). Assuming a system life expectancy of 30 years and that fossil-fuel-based energy was used in manufacture, 87–97 per cent of the energy that PV systems generate will be pollution-free. These figures were from 2004 or earlier, and so it is likely that they are shorter now.

Global warming impact

As with the rest of the semiconductor industry, many potentially dangerous chemicals are used in the production of solar cells. F-gases including sulphur hexafluoride (SF_6), nitrogen trifluoride (NF_3) and other climate-affecting and ozone-depleting fluorine compounds are being used in the industry and severely compromise the greenhouse gas balance of some PV modules. This technological area is young and fast moving, and processes and facilities may sometimes have a very limited economic lifetime. For example, the silicon thin-film industry is in the process of substituting SF_6 and NF_3 with climate-neutral gases. Some manufacturers are better than others, especially those in countries with rigorous environmental enforcement regimes.

A 2009 survey of the carbon impact of PV looked at the global warming impact of glass/glass-encapsulated thin-film amorphous silicon-based modules and nano-structured materials in thin-film silicon modules, used in rooftop-mounted modules installed in southern Europe, lasting 30 years (a comparatively rare type of module). It was found that the lifecycle impact of amorphous PV systems was between 0.03–0.04kg CO_2/kWh, and for micromorph thin-film 0.06–0.08kg CO_2/kWh. This compares favourably with the impact of natural gas at 0.195kg CO_2/kWh and electricity from the (UK) national grid at 0.55kg CO_2/kWh (based on the annual average).[11]

Pollution during manufacture and disposal

Many toxic chemicals are used during the manufacture of PV cells. A report by the Silicon Valley Toxics Coalition (SVTC)[12] says that there may be no cause for alarm about their impact in countries with well regulated environmental pollution regimes, but such strict monitoring and controls are not necessarily in force everywhere. It draws attention to the problems posed at the end of the panel's lives on their recyclability or disposal. SVTC is calling for solar panels to be classed as electrical and electronic equipment waste and to be subject to mandatory take-back and recycling legislation.

SVTC also compiles a 'Solar Company Survey and Scorecard',[13] which measures solar industry companies' commitment to take responsibility for the impact of their products. Not all manufacturers have responded to questionnaires asking for disclosure about their work practices. Nevertheless, it is worth consulting if you wish to buy responsibly. There is also pressure on manufacturers to organize take-back schemes so that modules at the end of their lives can be processed properly. At present, outside of Europe, where they fall under the Waste Electrical and Electronic Equipment Directive, legal requirements for responsible disposal are very often lacking.

Figure 6.48 Dye-sensitive solar cells can be printed on flexible transparent plastic and many other materials.

Source: ISE at www.exposolar.org/2010/eng/center/contents.asp?idx=88&page=1&search=&searchstring=&news_type=C

New PV technologies

Dye-sensitized solar cells

This type of cell was invented by Michael Grätzel and Brian O'Regan at the École Polytechnique Fédérale de Lausanne in 1991, and are also known as 'Grätzel cells'. They are made of low-cost materials and do not need elaborate apparatus to manufacture. The cells have a simple structure that consists of two electrodes and an iodide-containing electrolyte. One electrode is dye-absorbed highly porous nanocrystalline titanium dioxide (nc-TiO_2) deposited on a transparent electrically conducting substrate. The other is a transparent, electrically conducting substrate. TiO_2 is a very common substance, used in toothpaste and sun lotion.

The dye can be organic, of plant origin, like the colouring found in pokeberries or blackberries. When light falls onto the dye-sensitized solar cell, it is absorbed by the dye. The electrons that are excited, due to the extra energy the light

Figure 6.49 David Carroll, the director of Wake Forest's Center for Nanotechnology and Molecular Materials, with his new fibre cell using nanotechnology.

Source: Wake Forest's Center for Nanotechnology and Molecular Materials

provides, are able to escape from the dye and into the TiO_2, diffusing through the TiO_2 to the electrode. They are eventually returned to the dye through the electrolyte. Their lab efficiency is currently around 7 per cent, but it is hoped this will improve. They are cheap to produce and are predicted to be mainstream in a few years.

Fibre-based modules

Wake Forest University's Center for Nanotechnology and Molecular Materials in North Carolina, US, holds the patent for a fibre-based PV cell using dye-sensitized cell technology. Since the fibres create much more surface area, these cells can collect light at any angle – from sunrise until sunset. The plastic fibres are stamped onto plastic sheets, using the same technology that attaches the tops of soft-drink cans. The absorber – either a polymer or a less-expensive dye – is sprayed on. The plastic modules are lightweight and flexible – the inventor believes they can be rolled up and shipped cheaply to developing countries. Then, workers at local plants could spray them with the dye, which would be collected from local pokeberry plants, for example, and prepare them for installation.

Laminates with heating and cooling capability

PV laminates convert sunlight into electricity and heat at the same time. They are currently being researched in building-integrated applications and involve a combination of amorphous silicon PV laminates, phase-change materials used as a heat sink, and metal roofing panels, largely for industrial or corporate buildings. Initial results of the research are promising, showing that a relatively high solar absorption can occur in the winter for heating purposes, with the phase-change materials providing heat storage capacity and increasing overall attic air temperature during the night. Of course, they are also producing electricity. To compensate for the fact that the laminates may generate increased cooling loads in the cooling season, the heat sink is designed to minimize the summer heat gains. They can be retrofitted over existing roofs, which is a cost-effective way to repair them without generating waste, but the temperature of the phase change needs to be calibrated according to the conditions.

Nanotechnology and cell-printing

Conventional thin-film solar cells are made by depositing the dye in an expensive, vacuum-based process. Wouldn't it be much better if the dye could be printed onto any material using conventional printing technology? That is the goal of this technique, which utilizes nanotechnology. Solar power-generating surfaces could then be produced much faster and more economically. Fully automated factories are already printing, laminating and shipping such panels. They use panels made with copper indium gallium diselenide and nano-structured components as the basis for creating printable semiconductors. As demand rises, prices will come down.

Notes

1 E. Rudkin, J. Thornycroft and S. Matthews, *PV Large-Scale Building Integrated Field Trial Second Technical Report – Monitoring Phase*, UK Department for Business, Enterprise and Regulatory Reform, London, 2007.
2 B. Croxford and K. Scott, Can PV or Solar Thermal Systems Be Cost Effective? PhD thesis, London, 2008.
3 SolarBuzz, www.solarbuzz.com/ModulePrices.htm, accessed May 2010.
4 Arnulf Jäger-Waldau, *PV Status Report 2008*, Renewable Energy Unit, European Commission, Italy, September 2008.
5 Potential of On-Grid Photovoltaic Solar Energy in Sunbelt Countries by the Alliance for Rural Electrification, Brussels, December 2009.
6 Trends in the Global Solar Photovoltaics Industry, Aruvian's Research, 2010, and Marketbuzz® 2010 report, San Francisco.
7 Photovoltaics in 2010, European Photovoltaic Industry Association, Brussels, 1996.
8 Jäger-Waldau, *PV Status Report 2009*, note 4.
9 Solar Energy Industry Association (SEIA), US Solar Industry Year in Review 2008, www.seia.org, accessed May 2010.
10 NREL, 2004, What is the energy payback for PV? National Renewable Energy Laboratory, Washington, US, www.nrel.gov/docs/fy05osti/37322.pdf, accessed April 2010.
11 E. Alsema, and R. van der Meulen, 'Fluoride gas emissions from amorphous and micromorphous silicon solar cell production: Emission estimates and LCA results', Copernicus Institute/Utrecht University, the Netherlands, 2009. Available at: opus.kobv.de/zlb/volltexte/2010/8346/pdf/3889.pdf, accessed October 2010.
12 D. Mulvaney et al, 'Towards a just and sustainable solar energy industry', Silicon Valley Toxics Coalition (SVTC), San Jose, CA, US, 2009.
13 www.solarscorecard.com, accessed October 2010.

Reference

Martin, C. L. and Goswami, D. Y. (2010) *Solar Energy Pocket Reference*, International Solar Energy Society, Earthscan, London

7 Stand-Alone Photovoltaics

Off-grid PV systems

It is in remote and rural locations where solar electricity is already cost-effective and changing lives, as Hermann Scheer, discussed in Chapter 6, acknowledges. Where there is no connection to a national network, the magic of electricity is being made available using solar energy, as an alternative, or in addition to, diesel generators. The sun is being put to use to transform lives by facilitating fresh water through pumps, enabling valuable drugs to be kept in refrigerators, or supplying evening light so children can study after dark. PV modules in stand-alone applications are performing these – and more – functions. A partial list would have to include:

Worldwide:

- portable appliances – phones, radios, TVs, laptops;
- caravans, boats and ships;
- telecoms – e.g. repeater stations, mobile phone networks;
- street lighting, street furniture and signs;
- navigation buoys, security lighting, alarm systems;
- vehicle battery trickle chargers;
- desalination.

Developing countries and remote locations:

- home systems;
- lighting;
- clinics and hospitals;
- schools;
- rural workshops and businesses;
- water pumping;
- back-up for other generators;
- mini-grids;
- tourism (lodges, parks);
- two-way radio;
- emergency power for disaster relief.

We will look in more detail at some of these applications, but it is beyond the scope of this title to examine in a detailed way how stand-alone systems might be sized, specified and installed. That topic is covered in our sister title, *Stand-alone Solar Electric Systems*.

Figure 7.1 Independence from power cuts is one of the advantages of PV power.

Source: Steca

Figure 7.2 Stand-alone systems like this mean children in rural areas can study when it gets dark.

Source: Steca

Figure 7.3 Solar lantern using 32 LEDs, which give a long life of around 100,000 lighting hours, compared with halogen lamps which provide around 1000 hours, and U-type lamps giving around 5000 hours. The solar module is rated at 2.5W and the lamp has an integral battery with a capacity of 2.5Ah.

Source: © Dulas Ltd

There is no single market for PV globally, but a conglomeration of regional markets and special applications for which PV offers the most cost-effective solution. In the developed world, grid-connected solar systems outnumber stand-alone systems; in the US, for example, grid-connected systems overtook stand-alone ones in 2005. But in developing countries, the converse is true; in India at the end of 2008, most PV applications were off-grid – mainly solar lanterns, solar home systems, solar street lights and water pumping systems. Only 33 sites were grid-connected, with a total capacity of approximately 2MWp. India's Ministry of New and Renewable Energy, not to mention the Chinese government, have targets to increase the number of their countries' grid-connected systems with subsidy and power purchase programmes.

It is true that PV modules and their accompanying systems can seem expensive. They are, however, cost-effective when considered over the long term, including the total lifecycle costs, when compared to diesel generators, which require the constant purchase of fuel and sending regular maintenance into remote areas. PV panels require less maintenance, as they have no moving parts, and can last for 15–20 years. Used properly, their power output is largely predictable. They are reliable, if installed and used correctly and sufficient storage capability is provided. Training local technicians is crucial.

Portable appliances

The ability to charge battery-powered gadgets and appliances from sunshine is extremely valuable for preserving the continuity of electrical supply when away from a grid. Many types of appliance are now on the market that come readily adapted for solar power. Some have integrated solar panels, like radios, televisions and lamps. Most of the lamps are LED lamps; LEDs, or light emitting diodes, use extremely low amounts of electricity and are very effective for directional light. For example, for 240 lumens of light output over six hours, 22W is sufficient, which a solar module can supply with an output of 30W. Wind-up radios and lamps also come with solar panels on the top.

For appliances that consume more power, such as laptops, televisions and fridges, specially tailored modules supplying the correctly modulated current are available. For hikers, flexible panels can charge electrical equipment while they are walking. For boats, caravans and yachts, combined micro-wind turbine and PV modules systems are available with batteries for storage. That way, power can be supplied both when the wind is blowing and when the sun is shining. Wind power and solar power can make ideal companions in a so-called hybrid system because often when the sun isn't shining, the wind is blowing and vice versa.

In this context, the modular aspect of PV power comes into its own. It is quite common for systems to start small, and grow as budget allows, or as loads are added. (A 'load' is shorthand for 'electrical load' or any device which consumes electricity that is added to a circuit.) PV modules can be added to a system to increase the amount of power available. The simplicity of this, not to mention its user-friendliness, cannot be overstated.

Telecoms and weather stations

It is for their reliability and low maintenance requirements that PV is often chosen to power remote repeater stations and other components of a mobile phone network. (A repeater is a device that picks up, amplifies and transmits cellphone signals to and from the nearest base station.) Weather monitoring stations are another popular application. After all, there is some elegance about a weather monitoring station powered by the weather. Power supplies for these are frequently combined with wind turbines and backup diesel generators.

For example, in Hilaire, in northeastern Haiti, accessible only by a four-hour ride on horseback, is a telecoms tower at the summit of a mountain that rises to 914.4m (3000ft). A PV-battery system gives it completely autonomous operation, with a generator for back-up in the winter. In the mainland US, at a similarly remote location, Cedar Mountain, one of the highest peaks in Pennsylvania's Tioga State Forest, sits a microwave repeater station. The cost of extending utility power to this location over a distance of 11.3km (7 miles) was prohibitive compared to the cost of a solar/wind/generator hybrid system. This system operates autonomously from solar/wind power during summer. During winter, it operates approximately 65 per cent of the time from the generator and 35 per cent from the solar/wind combination.

There are plenty of examples of fuel cell power systems that supply back-up power to a remote PV-powered radio-telephone repeater or microwave relay station, where reliability is paramount. In this type of system, the fuel cell starts automatically when solar insolation is insufficient to maintain the state-of-charge of the system's battery. A cellular modem commonly permits remote monitoring and control of the system.

Street furniture

The same benefits make PV popular for street furniture, such as lights (whether illuminating traffic signs or bus stops), parking meters or traffic lights. Although there might be grid-sourced electricity in such an urban situation, it can make sense for local authorities to use a PV system because it provides an all in one turnkey solution without having to dig up the roads to make the connection. It also means that in the event of a grid power cut, the traffic system will not be disrupted. Often the installation comes with a service agreement.

Figure 7.4 A bus shelter in Berlin, Germany, with PV-powered lighting, and a PV-powered parking meter in Rome, Italy.

Source: © Frank Jackson

Figure 7.5 A speed warning system powered by PV in Wales.

Water services

Pumping water

Many parts of the world do not have a reliable supply of drinking water; water scarcity is an increasing problem. This can seriously affect people's health. However, solar power can be used to pump water from boreholes in situations where demand is too high or the well is too deep for the use of a hand pump. It is an ideal solution because systems can be easily sized to meet demand and are reliable – the only moving parts are in the pump itself; if that breaks down, it is relatively easy to repair. Other forms of power supply, such as a wind turbine or diesel generator, require higher levels of maintenance. Nevertheless, diesel can usually deliver water in greater quantities and from greater depths than PV, while PV is good for small- to medium-scale solutions where there is sufficient sunshine. Mechanical windmills can function very well; electrical ones often suffer because of the extreme conditions – high winds and dust can cause parts to wear out quickly. However, wind power may be cheaper where the wind resource is good. PV can be combined with either of these if required. Solar electricity is not normally cost effective for use with deep boreholes, such as those over 150m (500ft).

The relatively high initial cost of the modules (perhaps three times the cost of a diesel generator) is offset by considerable lifetime savings on fuel and reduced maintenance expenditure, and will soon make up for this initial outlay. The systems are ideal for meeting drinking water requirements for villages of up to 2000 inhabitants. The pumped water can also be used for animal watering crop irrigation (up to 3ha or 7.5 acres in scale, at low heads (vertical heights), depending on water requirements).

The water is pumped when the sun is shining and stored in an elevated tank for use when there is no power. The system is consequently relatively simple; it operates on DC, and an inverter is not required. The tank must be appropriately sized to store sufficient water for overnight and cloudy periods. The water in the tank is able to be gravity-fed to the taps or watering points for livestock, or the irrigation system, provided that the tank is located at least 30m (100ft) above the homes to be supplied; this will provide the necessary 20psi pressure. The only maintenance required for the modules is regular cleaning, but the modules must be placed in a secure position. A 1kWp array will, on clear sunny days in hot regions of the world, pump the following volume of water, the following vertical distance, per day:

volume	height
50m^3	10m
15m^3	30m
10m^3	50m

Adding further kilowatts simply multiplies the volume or the distance linearly. For example, a 2kWp system will pump enough water to supply a community of about 1400 people. However, this is a rough guide only and it is crucial that the system is correctly sized to the local conditions.

There are different types of pumps for different applications, such as submersible and surface pumps. It is important to match the components in a solar pumping

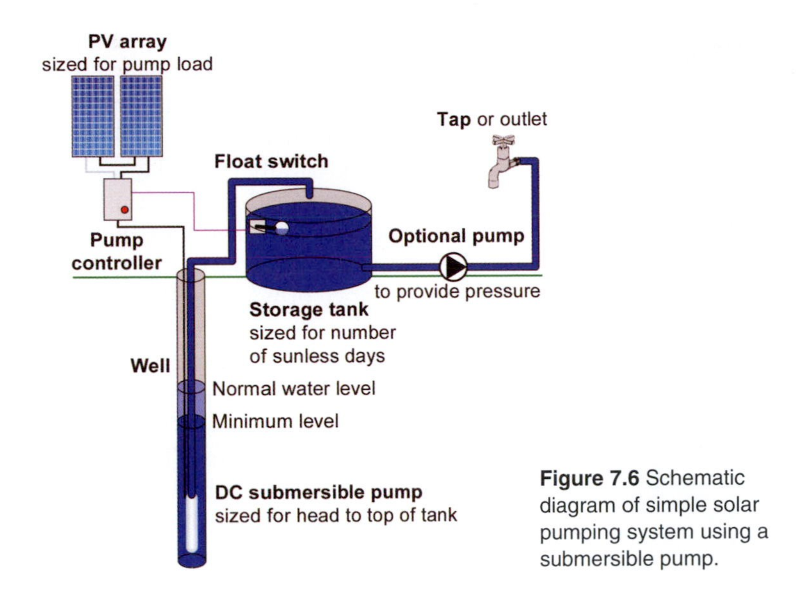

Figure 7.6 Schematic diagram of simple solar pumping system using a submersible pump.

system. For example, the correct choice of pump and motor depends on the quantity of water and the depth from which it is to be pumped. Normally the entire system is purchased as a kit, and selected once the local insolation figures, type of pump, daily water requirement and the head (vertical height) are known.

Desalination or distillation

PV can also be used to provide drinking water in another way, through the desalination of water. In this application it is also powering a pump, but this time to push the water through a reverse osmosis (RO) membrane. RO is a pressure driven process, where pure water is continuously drawn from salty water through a semi-permeable membrane. Again, PV is an alternative to using diesel or wind power, and is suitable for small-scale plants.

In a typical system, a PV cell array is used with or without a battery bank to power a DC pump and push the fresh water into a storage tank. Such a system will include a cartridge pre-filter and a spirally wound RO membrane element. In sun-belt areas, for 15W input, 0.1m^3 of fresh water per day can be obtained, which can cater for the drinking and cooking requirements of a couple of families.

Some companies supply bespoke solutions for water purification and desalination. These are small-scale devices, with the module typically 50–100W, with built-in batteries. A unit consists of a complete kit containing a solar module and a water purification or desalination unit (they are different products). A water purification unit will contain water pre-filters, an ultraviolet disinfection lamp and, say, a 100-litre (22-gallon) water tank. Relatively portable, they might weigh 40–50kg. The purification units are effective against viruses, bacteria, protozoa and worm eggs. Desalination units will remove dissolved salts. In clear tropical conditions, a 75Wp system can provide around 2000 litres (440 gallons) – enough for 300–400 people.

More sophisticated systems use AC pumps and ultraviolet treatment to disinfect the water and provide greater reliability, which requires the use of an

Figure 7.7 Dutch company Nedap's all-in-one solution for PV-powered water purification.

Source: © Nedap NV

inverter and means the array must be larger for the same volume output. Batteries are included to increase reliability, since fresh water is a vital supply. An example of such a system is in use at Trombay, India. In a tropical region, using this equipment, 2m³ (20ft³) per day can be supplied by a 2kWp array. In subtropical regions, this figure could drop to 0.8m³ (28ft³) per day, which is clearly not cost-competitive. In any region, economic viability will depend upon the availability and price of water, and the existence of subsidies or relative fuel price.

Another, community-scale example has been in use since 2006 in Rajasthan, India, on the shore of the Sambhar salt lake. It has been producing 5000 litres (1100 gallons) per six hour day from a 3kW array. The water comes from the salt lake, whose water is otherwise undrinkable, around which 100 villages produce salt by evaporating the water. This system makes both salt and water.

A second method is membrane distillation (MD). There are two differences between MD and RO. One is in the pore diameter of the membrane used. For RO, membranes have a pore diameter in the range of 0.1–3.5nm; membranes for MD generally have a pore diameter of 0.1–0.4 microns. At this much smaller size, up to a certain limiting pressure, surface tension prevents water from entering the pores, but water in the form of vapour can pass through the membrane, to condense in a pure form on the other side, where it is collected.

The other difference is that a solar thermal collector is used to raise the temperature of the water to 80–85°C (176–185°F), before it is introduced to the membrane by the PV-powered pump. A basic description of this system is that the salty or polluted feed water is pumped to the condenser inlet of the membrane module, where it is pre-heated and enters the solar thermal collector at the bottom. It rises, warms, and is pumped into the evaporator of the membrane module. Its heat is communicated to the incoming water, the distillate is collected and any residue returned to the brine tank for recirculation. Up to 150 litres (33 gallons) per day can be produced by 6.7m² (72ft²) of solar thermal collectors and a small PV-driven pump.

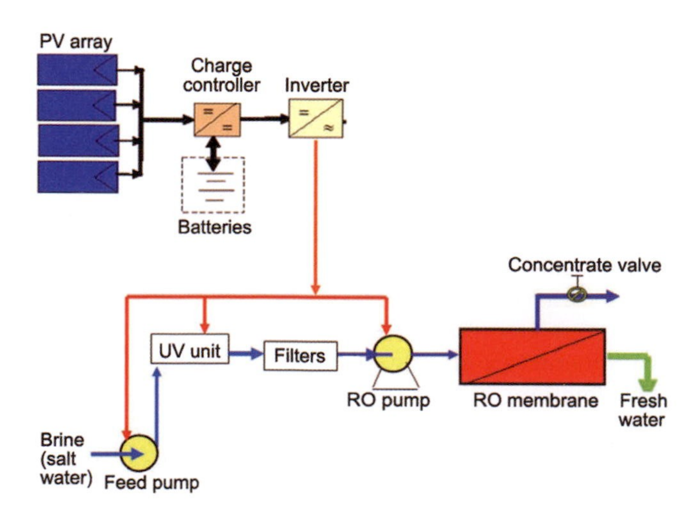

Figure 7.8 Schematic diagram of an AC desalination system.

Refrigeration

Fridges are used by health agencies in rural health centres in remote areas to store blood banks and vaccines. When powered

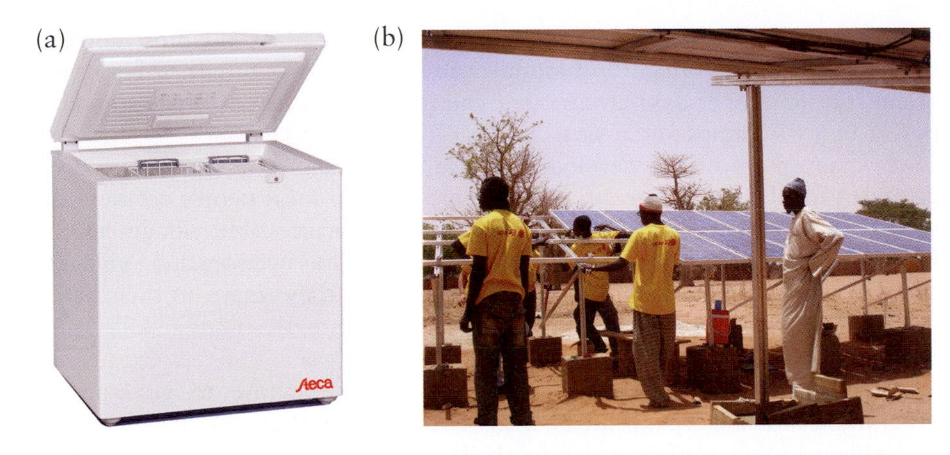

Figure 7.9 (a) The Steca low-energy 12/24V fridge/freezer has been developed for use in off-grid solar systems – its low energy rating (A+++) means that in most parts of the world it can be powered with a single 75Wp module. (b) Village mini-grid being installed in Senegal by SOLAR23 and local partners.

Source: (a) Steca; (b) SOLAR23 GmbH at www.solar23.com

by a PV system with battery back-up they provide a reliable service in countries as disparate as Yemen, Indonesia, Ghana and Nicaragua. Larger, turnkey systems, which include a fridge whose power requirements are perfectly matched with the PV system, are available for aid organizations and emergency relief. These typically include low maintenance, sealed, high-performance batteries (non-hazardous), an inverter for powering AC equipment and a charge controller with an easy-to-read display.

Home systems

In many rural areas, solar home systems provide power for a few small fluorescent lights and other small appliances. For example, a survey conducted in Sri Lanka by the International Resources Group shows that most of the solar home systems use a 50W PV/battery system to power several compact fluorescent lights.[1]

Basic system design

A typical, small off-grid or stand-alone system will include the following components:

- a PV module or array;
- a battery charge controller;
- batteries;
- safety disconnects and fuses;
- cables;
- an inverter or power control unit, if it is necessary to convert DC to AC (in larger systems);
- loads: e.g. fluorescent lamps (usually DC) rated 6–11W.

The PV modules are commonly rated 20–80Wp with 50Wp the most popular size. A system based on a 20Wp module can supply two or three 6W lamps for about four hours per day: at the other end of the range, an 80Wp system can power four 8W lamps and a small television set.

In many parts of the developing world, training programmes have been going on for the last 10 or 20 years to give skills to local people so they can install such systems themselves. For example, in inaccessible villages in the Himalayas, the Barefoot College in Rajasthan, India, has been teaching villagers to be 'Barefoot Solar Engineers'. After the training, they return to their home villages to install solar units and provide their communities with a skilled and competent repair and maintenance service.

Such systems are spreading in popularity in response to government programmes and aid organizations. In 2006, the Aryavart Gramin Bank installed PV systems in five of its branches to charge the back-up batteries that provide AC power in the event of a power cut from the mains. The bank then realized that such systems would be ideal for the needs of many of its rural customers, who had no access to electricity. It initiated a programme that helped dealers in local villages buy bulk orders of solar home systems and provided them with commercial loans. The systems support lights, a mobile phone charger, a DC fan and/or a basic television.

In many places, suppliers nowadays offer complete systems sized to particular specifications. They will calculate the optimum appropriate components. It is worth obtaining several quotes from different dealers, and ensuring that they have the appropriate certification, training and experience. They will need to know the latitude and location of the site and preferably the amount of insolation it receives year-round – or they will have access to such data.

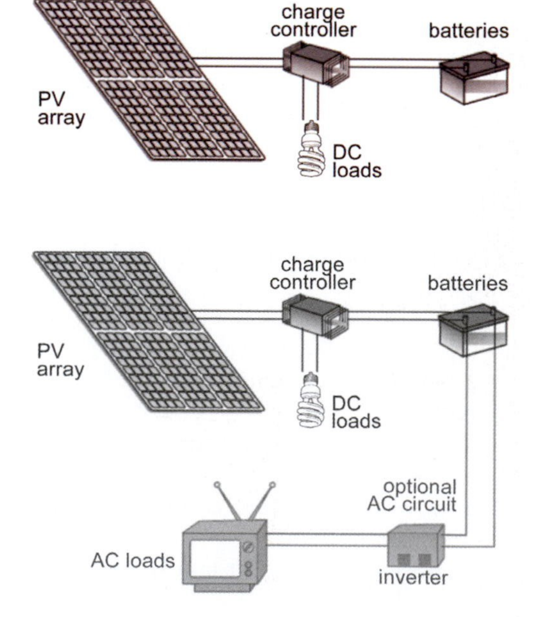

Figure 7.10a A basic circuit for a home system will include a charge controller, batteries and loads, typically lights and perhaps a radio. The loads will be DC.

Figure 7.10b If the system is to power AC equipment then an inverter is required.

The system components that are ideal for one site or purpose may not be optimum for another, so it is worth checking that the supplier does not supply a 'one-size-fits-all' solution. They should be able to demonstrate the reasons for their recommendations and technical documentation. A reputable dealer will offer a warranty and service agreement. Correct installation and user-training is essential.

Evaluating the site

Not all sites are suitable for solar power. The module(s) will need to face the equator (south in the northern hemisphere, north in the southern hemisphere), and ideally receive sunlight throughout the whole of the day all year round. It should not be shaded by hills, buildings or trees, especially between the peak sun hours of 10am to 3pm. There should be sufficient space to mount the required number of modules where they will not be at risk from damage or overheating – efficiency tails off significantly at higher module temperatures. The roof or mount for the panels must be able to withstand their weight, wind forces, plus the weight of any snow that might fall during winter months.

Reducing demand

The first step in designing the system is to work out how much power is required. This usually involves minimizing the power requirements, since it is far cheaper to install efficient appliances – fridges, lighting, PCs and monitors, for example – than to buy a system to power hungry loads. Small PV systems are not practical for heating space or water, or air conditioning, electric stoves or electric clothes dryers, which all use large amounts of energy. If pumped

Figure 7.11 Solar training courses in action in Tanzania (left) and Abuja, Nigeria (right).

Source: © Frank Jackson

water is required for toilets and showers, then the most water-conserving cisterns, taps, showerheads and low bore pipes should be installed, to minimize the work that needs to be done by the pump.

Batteries

Batteries are an essential part of an off-grid system, ensuring its reliability when the sun isn't shining. The choice of the right battery is therefore essential. Battery care is something of an art, and probably provides the steepest learning curve for any owner of such a system. Batteries that are not properly looked after can severely reduce the efficiency of the entire system. Over the lifetime of the system, they may even have the highest cost, as they need to be replaced from time to time. Some batteries are sealed and require less maintenance, but all of them wear out and need a full charge regularly.

Batteries used in PV systems are mostly lead-acid batteries. These fall into two types: 'deep discharge' and 'shallow discharge'. The former are preferred because they can be almost completely drained without too much damage. Auto batteries are of the shallow discharge type and should be avoided. However, lead-acid batteries are heavy. Other types, such as nickel-metal-hydride, nickel-cadmium and lithium-ion batteries, are sometimes used because they are lighter, require less maintenance, have a longer life and are more flexible in use, but the trade-off is a usually unacceptable higher price. Fuel cells are used in high-performance situations.

The battery bank for a given system is chosen according to its capacity – the maximum amount of storage time required for a given amount of electricity that would be used during that time. This period of time is usually a reasonable number of cloudy days, when the batteries cannot be charged. The unit used to describe battery capacity is amp-hours (Ah); this signifies the amount of energy that can be drawn from the battery before it is completely discharged. For example, 100Ah can deliver 1A for 100 hours, 2A for 50 hours (though in fact it will be less because amp-hour capacity is dependent on discharge rate), etc. The amount of current needed depends on the loads expected to be required during that period. Once calculated, this figure is used to purchase the batteries. For example, if the requirement is for 506Wh for three days, divide by the voltage of the system (usually 12V) and multiply by the number of days (3):

$$506/12 \times 3 = 42 \times 3 = 126\text{Ah}$$

Batteries come in sizes defined by their capacity in amp-hours. To store 126Ah, then, we might buy three deep-cycle batteries of 56Ah capacity each – extra capacity is always needed as batteries should never be completely discharged. Expert advice should always be sought.

System suppliers can usually recommend the particular type of batteries that work best with their modules, since not all batteries are the same. Lead-acid batteries give off explosive hydrogen gas when charging. They should therefore be housed in a well-ventilated space away from the other electrical system components and living spaces. Follow the manufacturer's maintenance instructions and recycle the batteries when they wear out.

Charge controllers

Charge controllers include a low-voltage disconnect that prevents over-discharging, which can permanently damage the batteries. They also extend battery life by regulating the flow of electricity to keep the batteries fully charged without overcharging, and protect the battery bank from overcharging. It monitors the charge level in the battery bank and when it senses that the bank is fully charged it interrupts the flow of electricity from the PV panels. Some charge controllers also include maximum power point tracking (MPPT); this optimizes the PV array's output, increasing the energy it produces. The choice of controller depends also on the size of the PV system to which it must be matched, and the system voltage.

Maintenance

Modules should be inspected and cleaned regularly. The power output should be monitored and recorded – this creates an understanding of the system and an ability to spot possible issues. The electrolyte level of the batteries should be kept topped up, and the system should tell you how much charge is in the battery at a given time. If this is not done automatically, batteries can be read with a meter or a hydrometer. It is also important to regularly keep an eye on the PV array voltage, the module and battery temperature, and perhaps inverter performance.

AC systems

In addition to the above components, an AC system needs an inverter to convert the direct current (DC) to the alternating current (AC), which powers standard mains equipment. This will is installed between the batteries and the AC loads. The inverter needs to be of the 'stand-alone' type (not one for feeding power into the grid, as in grid-connected systems). Inverters vary according to the voltage and current they can accept and the frequency and power of the output. Therefore one should be selected to match the peak load it will have to meet; at the domestic scale they range from 100W to 10kW. The quality of output (waveform) also varies with cost: sensitive electronic equipment needs a smoother supply – a sine wave, which is like normal mains power – than lighting, TVs and power tools, which can accommodate a square wave output. The latter are generally cheaper than the former.

Inverters consume energy, and this wasted energy can be a high proportion of what is generated. Therefore, it is preferable if at all possible to avoid the need for an inverter by choosing DC versions of appliances and lighting, especially in smaller systems. An inverter's efficiency is expressed as a percentage ratio of power out divided by power in, and it varies continually with the operating conditions: the load demand, power factor and input voltage. If an inverter is required, then the maximum losses should be factored in when designing the system and deciding how many modules to install. Inverters should always be located in a cool, well ventilated, dry place.

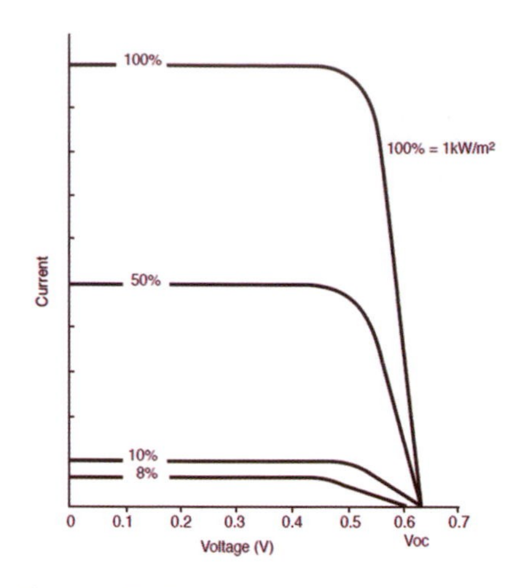

Figure 7.12 The amount of voltage and current produced by a polycrystalline cell at different light intensities. Voltage is affected far less than current, which means batteries can be charged even on a dull day.

Source: RS Components

Hybrid systems

As the load requirements increase, the addition of another type of generator to a PV power supply system often makes sense. Configurations typically include one or more renewable energy source (PV, wind, etc.), battery storage, and a diesel generator for back-up. Hybrid systems have all the sources charging a single battery bank, with each of them requiring its own charge controller. The loads are then powered from the battery bank.

A PV/diesel hybrid system is frequently a viable option, even with high fuel costs. The load for a small local grid supplying a village may include a community centre, schools and workshops. Adding a back-up diesel generator to such a system serves peak demands and provides power when the sun isn't shining. It permits smaller PV modules and batteries, reducing system costs, while the PV and batteries reduce the amount of diesel fuel needed. For example, a 20kW PV/75kW diesel hybrid system in Gaize, Tibet, meets a daily energy consumption of 75kWh for a village of 1000 people. Equally, a wind turbine or two might be included.

But how are we to determine at what point their addition becomes cost-effective as more loads are added to the power requirement demands? When making this calculation, it is important to compare the whole-system lifetime (or 'levelized') costs. Although a PV system may be more expensive initially than a diesel system, the cumulative costs begin to be cheaper after between five and ten years. Software is available to help with the calculations; for example

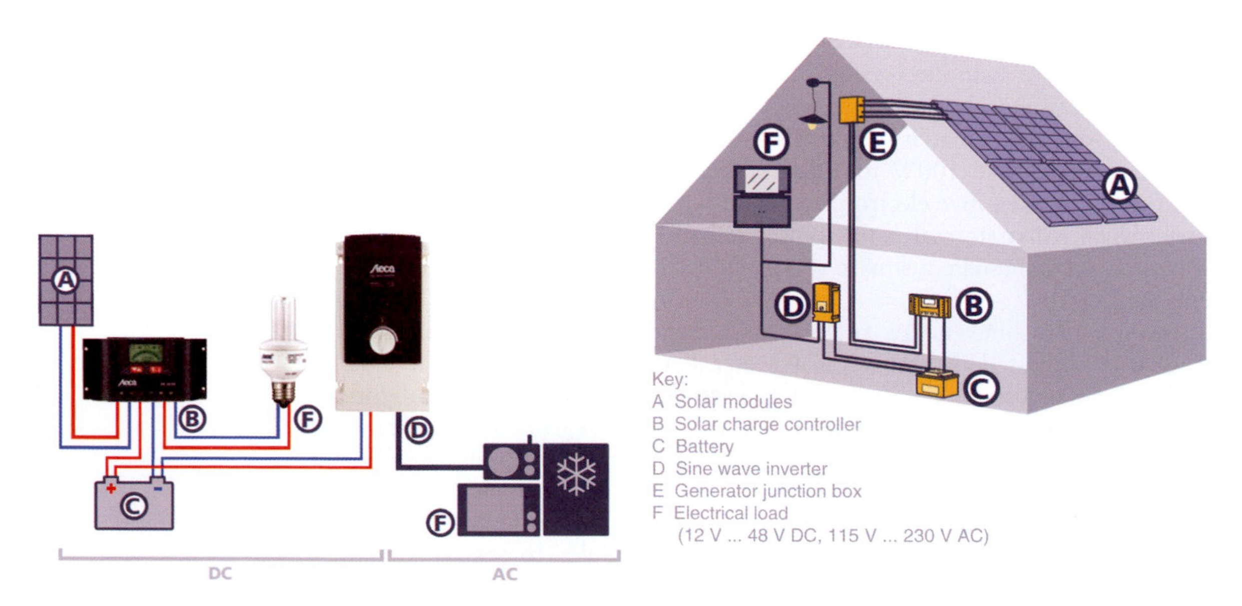

Figure 7.13 This system includes some loads that are DC and some that are AC and so requires an inverter.

Source: Steca

the HOMER software (see Chapter 10 Resources). Among the variables it can take into account are the local price of diesel fuel and batteries, and insolation data. Cost curves are then constructed that show at what points different options become viable. HOMER is useful in the planning and early decision-making phase of rural electrification projects. Further information on all these issues can be found in Chapter 10 Resources.

Mini-grids

Medium-size systems can be devised that operate as a 'mini-grid' for a village or cluster of workshops or other buildings. An array of panels and, quite often, wind turbines or other generation devices, will feed into the same system. Controllers, fuses boxes, meters, batteries and inverters will be housed in a dedicated shed. Cabling will then distribute the power to individual buildings. These complicated systems require careful installation, load matching, controls and monitoring. Greater efficiencies are found by matching particular inverters to particular sets of generators.

For example, at the Centre for Alternative Technology (latitude 52.6°N) in mid-Wales, many different PV devices, hydroelectric and wind turbines connect into the same system. At one point, it was a stand-alone system but around 15 years ago it became connected to the national grid. A PV array on the small roof

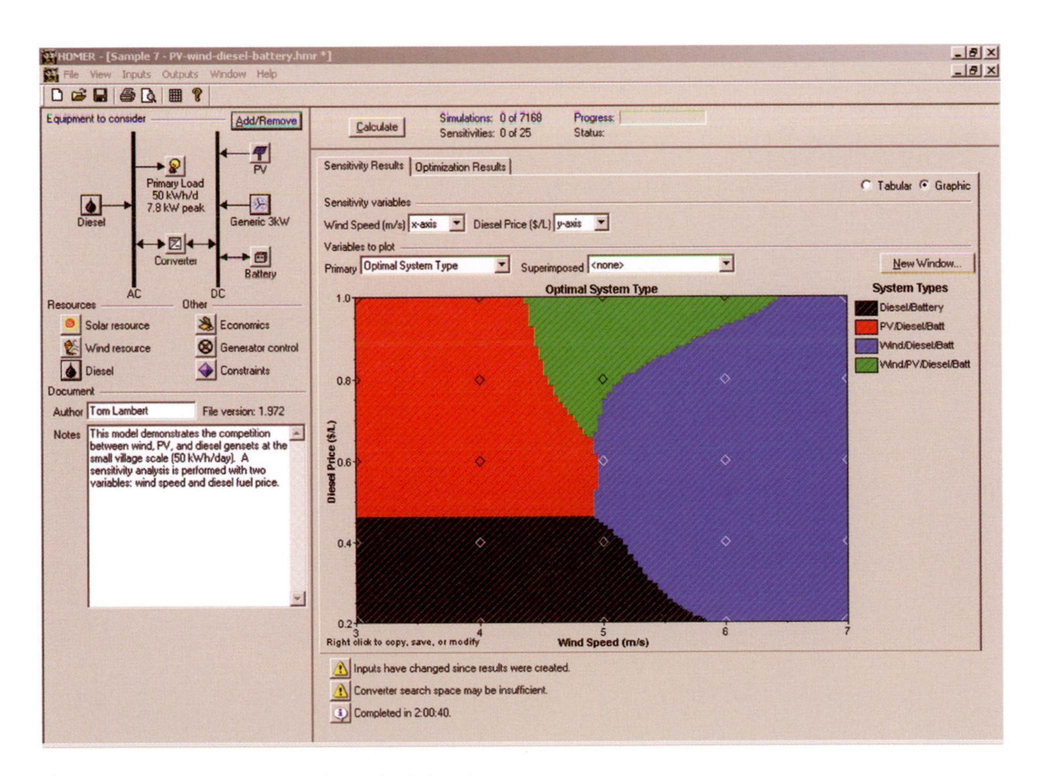

Figure 7.14 A screen grab of the HOMER software.

Source: HOMER

of a building housing a hydro turbine is rated at 1.68kWp. It has been monitored over the four years since its installation. In that time, it has averaged 1170kWh per year, or 696kWh/kWp/year. As well as the locally rather cloudy weather, it has two reasons to slightly under-perform – a south-easterly rather than a southern orientation, and the presence of a hill running along the south-easterly edge of the site. This truncates the winter days considerably.

The future

There's no doubt that over the next century PV has the power to transform the way in which we use energy, and power the gadgets we have come to regard as essential. But PV is not the only show in town that turns light into electricity. There are other technologies, one of which has been around for a lot longer, quickly emerging with huge potential. This we examine in the next chapter.

Note

1 International Resources Group, *World Bank/Sri Lanka Energy Services Delivery Project, Impacts Assessment and Lessons Learned*, Washington, DC, 2003.

8

Concentrating Solar Power

Concentrating solar power (CSP) systems work on a very different principle to PV. Instead of using light, CSP technologies capture the sun's heat rays to create mechanical power, which nowadays is usually converted to electrical power. They have been around for a long time, as the nineteenth-century illustration in Figure 8.1 shows. There is a variety of technologies for converting the heat into useful power, including lenses, mirror shapes, sun-tracking methods and arrangements, but they all work on the same principle. Like a child using a lens to burn a hole in paper, they focus or concentrate the sun's infrared frequencies, reaching temperatures of 200–1100°C (392–2012°F). This commonly gasifies a liquid to drive a motor, using the same principles as a steam engine or turbine.

CSP systems have some advantages over PV for large-scale use. They can be integrated with energy storage – using molten salts – so they can provide power when the sun is not shining. They can also be used in tandem with traditional fossil fuel or biofuel combustion to generate power, thereby guaranteeing capacity and reliability. However, having more moving parts than PV panels, they can be more costly to maintain. They are suitable for peak loads and base loads, and power is typically fed into an electricity grid. They are a proven technology that is currently causing much interest as a practical large-scale solution to the world's energy crisis. Many large plants are in planning or under construction in sun-belt areas.

Figure 8.1 The first parabolic trough – and the first utility-scale solar technology – from as long ago as 1883. John Ericsson's design focused sunlight from a curved silvered window glass surface (called a 'heliostat') onto an 3m (11ft) long iron tube central receiver to generate steam to mechanically drive a Stirling engine. The heliostat was 5m (16ft) wide and 3m (11ft) tall, and tracked the sun across a north–south axis. It had a maximum output of 2.24kW and was able to pump 2273 litres (500 gallons) of water per minute.

Source: © Creative Commons

History

In 1913, inspired by John Ericsson's pioneering work, American engineer and inventor Frank Shuman commissioned the first large-scale solar power generator in Maadi, near Cairo, Egypt. With a solar collector area of 1240 m^2, it powered a pump that irrigated elevated farmland with water from the River Nile. It consisted of five rows of parabolic mirrors with a total output of 88kW. This power station was more cost-effective than a similarly sized coal-based plant, and would have recouped its investment in four years. However, despite its success, it was only used for one year, as the First World War intervened.

This war began the worldwide predominance of oil as an energy source, as one of its major causes was the competition between the waning British empire and the emerging German economy over the Middle Eastern oilfields. By the end of the war, British forces had secured the entire oilfields of Mesopotamia in a new League Protectorate called Iraq. This put a provisional end to any attempts of pursuing the development of solar energy on a large scale.

Figure 8.2 The world's first solar power plant, in Egypt, 1913.

The pivotal importance of oil subsequently shaped the history of the 20th century and continues to be a major factor in global affairs. It is not only the basis of economic growth and energy-consuming technology, but has constantly been the cause of geopolitical tension and power struggles. From this perspective, the story of the power plant in Maadi, financed by London and built with Egyptian labour, has direct relevance to the present, not least the two recent Iraq wars.

Shuman's solar power generator wasn't the first use of concentrating solar energy. We read in the introduction of Augustin Mouchot and Abel Pifre's 19th-century solar-powered printing press. School teacher Mouchot's motivation was partly because he foresaw a time when coal supplies would be exhausted. Besides, the power source was free and easily available. But as far back as 672 BC the Chinese were using curved mirrors to start fires. In the 4th century BC, Euclid wrote about the same subject and, in 133 BC, Hero of Alexandria studied the phenomena of light reflection on flat mirrors, convex or concave. In Leonardo da Vinci's 1515 notebooks, we find a sketch of a CSP device. The first known such device to be built, using mirrors, was for the French king, by Georges Louis Leclerc de Buffon in 1747, which de Buffon called 'The Mirrors of Archimedes', after the Greek inventor's purported burning of Roman ships from a cliff top using curved, burnished shields while defending Syracuse in 212 BC. In 1767, the Swiss Horace Benedict de Saussure built various solar machines, which, in a performance improvement, were covered by glass on the side facing the sun to increase the greenhouse effect. In the late 18th century, the pioneering French chemist Antoine Lavoisier made the first solar oven, by concentrating sunlight using a liquid lens. It reached the then astonishing temperature of melting platinum, about 1800°C (3272°F). Sadly, this aristocrat's researches were curtailed when his life was ended by the guillotine during the French Revolution of 1789–1799.

Figure 8.3 The first parabolic solar collector, designed by Augustin Mouchot.

These pioneers paved the way for Mouchot, who singlehandedly advanced the solar cause in numerous ways. Here are a few of this visionary man's inventions:

1860: invented the first parabolic solar dish
1866: demonstrated his solar motor to Napoleon III in Paris
1876: made the first solar pasteurizer
1877: sent by the French government to French Algeria to carry out solar experiments in the desert
1878: constructed the first parabolic concentrator to power a thermal engine
1879: first solar production of ice
1879: first solar power generation of electricity.

Mouchot's assistant Abel Pifre developed the solar-powered printing press. His device consisted of a 3.5m diameter concave mirror focusing heat on a cylindrical steam boiler, which powered a small vertical engine of 0.4 horsepower that turned a Marioni type printing-press. The system operated even under semi-cloudy conditions.

Solar thermal was of especial interest in the second half of the Industrial Revolution, which was hungry for more and more energy. From 1868 to 1875, John Ericsson, the visionary Swedish American engineer responsible for many innovations in motors and power, designed the solar hub shown in Figure 8.5b, that was exhibited in New York, in 1872. In total, he built seven 'sun motors' powered by steam or hot air. In 1887, two years before his death, Ericsson wrote:

> *The time will come when Europe must stop her mills and factories for want of coal. Upper Egypt then, with her never-ceasing sun power, will invite the European manufacturer to move his machinery and erect his mills on the firm ground along the sides of the alluvial plain of the Nile, where sufficient power can be obtained to enable him to run more spindles than a hundred Manchesters.* (*Mechanic and Builder*, July 1887.)

Figure 8.4 Antoine Lavoisier's solar oven, which used a lens to focus the sun's rays, reached 1800°C (3272°F) in 1785. It rotated on a horizontal axis to track the vertical movement of the sun along its east-west axis, and was mobile.

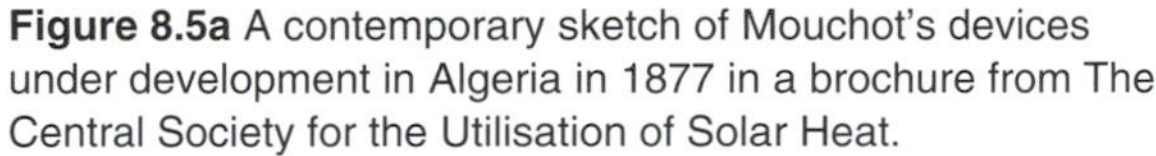
Figure 8.5a A contemporary sketch of Mouchot's devices under development in Algeria in 1877 in a brochure from The Central Society for the Utilisation of Solar Heat.

Figure 8.5b John Ericsson's 'sun motor' built for pumping purposes in New York in 1872.

This paved the way for the American Frank Shuman's Egyptian CSP station, described above. Shuman created the Sun Power Co., and dreamt of building 52,600km^2 of mirror collectors in the Sahara, which he calculated would produce 198MW – equivalent to the world's energy consumption in 1909. Many thinkers and developers are now reviving the dream of using the world's desert hot-spots to produce all the energy we need. As we shall see later in this chapter, perhaps its time has finally come.

Meanwhile, back in 1874 on the other side of the Atlantic, at Las Salinas on the high plateau of Atacama in Chile, a solar still produced 23,369 litres (5140 gallons) of fresh water each sunny day, and worked for 40 years. In 1901, in sunny Pasadena, California, a 10m (33ft) diameter solar dish in an ostrich farm pumped 5300 litres (1400 gallons) per minute. It did so by heating a 377-litre (83-gallon) boiler, and producing 3.36kW. Three years later, Mario António Gomez, a Portuguese and another major pioneer of solar energy, showcased the Pirelióforo, in which thousands of mirrors over a surface of 80m^2 (861ft^2) concentrated solar energy up to 3500°C (6332°F) – enough to melt most metals and stones. It won two gold medals at the 1904 Universal Expo of St Louis.

Finally, in the last historical development before modern CSP stations, another Frenchman, Felix Trombe, a chemical engineer, built a 2kW dish concentrator and then a 50kW solar furnace at Mont Louis, France, from 1946 to 1949. The installation still exists as a visitor centre. His intention was to reduce the need to burn wood for fuel. Trombe also invented the Trombe Wall (see the solar thermal space heating section).

Figure 8.6 Felix Trombe's solar furnace in Mont Luis, France.

How it works

In modern concentrating power station technology, the sun's heat is concentrated using (a) highly reflective surface(s) and sometimes lenses to a medium to high temperature. The troughs concentrate the incident solar radiation 80-fold. The heat is converted to mechanical and then electrical energy, using a steam or gas turbine, or a Stirling engine. The heat energy collected during the day is in some cases stored in liquid or solid media, such as molten salts, ceramics, water or concrete, and released at night to keep the turbines running. For example, the 49.9MWe Andasol plants in Spain have 209,664 mirrors and, besides running turbines, are hooked up to thermal storage, with an outlet temperature of about 290°C (554°F) and a top temperature of 390°C (734°F), increasing their annual availability by about 1000 to 2500 hours. The storage is achieved with 28,500 tonnes of molten nitrate and potassium salt storage that keeps it operational up to 7.5 hours after sunset, or during a cloudy period. Each plant has two tanks to hold the salts, built like Thermos flasks to preserve this temperature for several weeks. This means they can generate almost round the clock in the summer months. In order to charge the storage at the same time as operating the turbines, the solar field is significantly larger than it would otherwise be.

There are four main types of commercial CSP systems, in which the concentrating mirrors either focus on a line or a point:

- parabolic troughs and linear Fresnel systems are line-concentrating. They concentrate radiation about 100 times to achieve temperatures of up to 550°C (1022°F);
- central receivers (also called solar towers) and parabolic dishes are point-concentrating. They can reach temperatures over 1000°C (1832°F).

The main elements of a system are: concentrator, tracking system, a receiver, transport media or storage, and power conversion. Many types exist, including combinations with other renewable and non-renewable technologies.

Figure 8.7 The first parabolic trough power plant in Europe, the Andasol 1, near Guadix, Andalucía, Spain. It went online in March 2009. The site's high altitude (1100m; 3610ft) and the semi-arid climate, gives it exceptionally high annual direct insolation of 2144kWh/m² per year.

Source: © Wikimedia Commons

Parabolic trough

Parabolic trough collectors concentrate sunlight onto Dewar tubes (whose principle is similar to a Thermos flask) containing a fluid, such as synthetic thermal oil, direct steam or molten salt, situated in the trough's focal line. The troghs usually track the sun along one axis – east–west and are aligned north–south. The fluid can reach about 300–550°C (572–1022°F) and is pumped through heat exchangers to produce superheated steam. The steam drives a conventional steam turbine generator to produce electricity, or it can drive a combined steam and gas turbine cycle. Plants are typically 50–400MWe.

Applications

Most commonly, this technology is used in grid-connected power plants, generating mid to high-process heat. The highest single unit solar capacity to date is 80MWe. The total capacity built is over 500MW and more than 10GW are under construction or proposed.

Figure 8.8 (a) Andasol 2 power station, connected to the grid in 2009. Andasol 3 was due to be completed in 2011 as we went to press. (b) The principle of operation of the Andasol thermal power plants.

Source: © Solar Millennium AG 2008

(a)

(b)

The thermal storage system is loaded during the day.

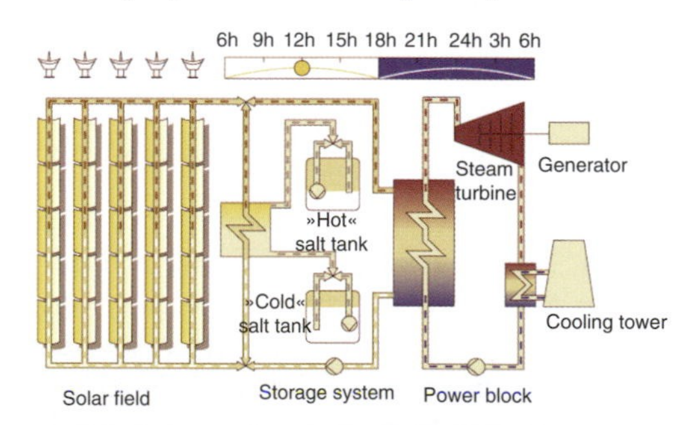

During the night, the power plant can be operated with the stored energy

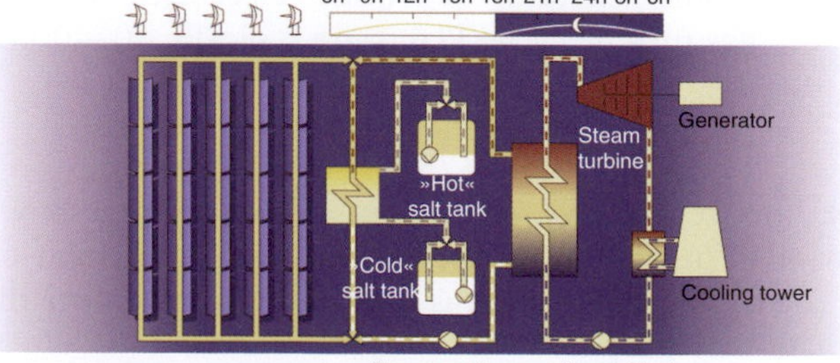

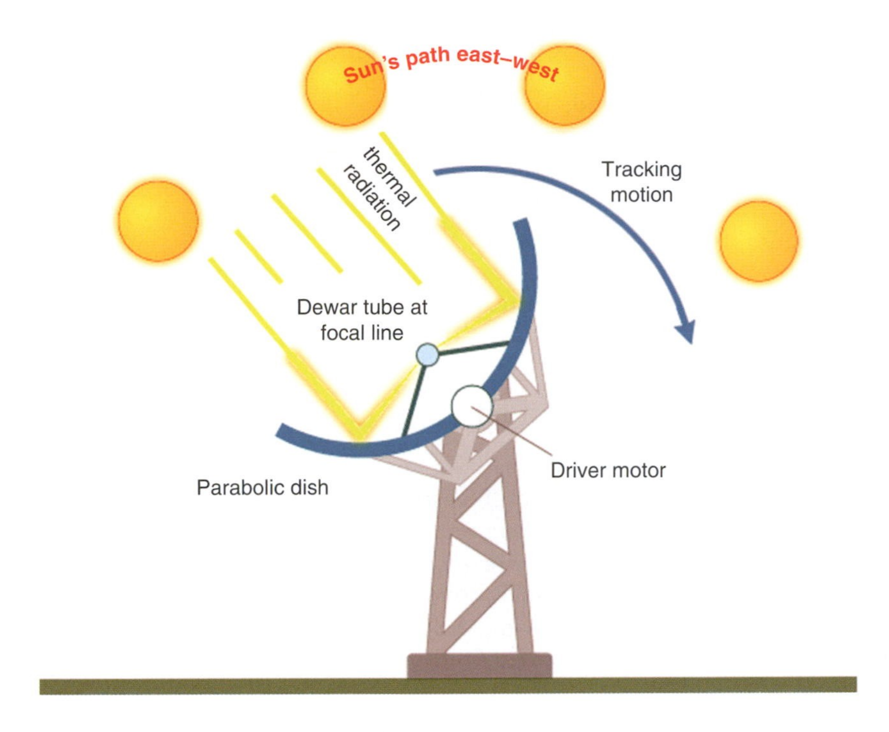

Figure 8.9 A parabolic trough has a tracking mechanism to follow the sun's path around the sky.

Figure 8.10 A parabolic trough solar thermal electric power plant located at Kramer Junction in California.

Source: © Wikimedia Commons

Advantages

- commercially available – over 16 billion kWh of operational experience; operating temperature potential up to 500°C (932°F) (400°C [752°F] commercially proven);
- commercially proven annual net plant efficiency of 14% (solar radiation to net electric output);

- modular;
- good land-use factor;
- lowest materials demand;
- hybrid concept proven;
- storage capability.

Disadvantages

- The use of oil-based heat transfer media restricts operating temperatures today to 400°C (750°F), resulting in only moderate steam qualities.

Linear Fresnel Reflector (LFR)

Rows of flat or slightly concave, long mirrors, made on the Fresnel lens principle (which is cheaper than real lenses and parabolic mirrors) concentrate solar heat onto central, convex, long receivers located at the focal point. Water flowing through the receiver is turned into steam at 200–500°C (390–930°F). This approach is similar to the parabolic trough but it has lower costs for its structural support and reflectors, fixed fluid joints and its receiver – separated from the reflector system. Existing plants vary in size from 5 to 400MWe.

Applications

LFR systems are suitable for grid-connected plants, or steam generation to be used in conventional thermal power plants. The highest single unit solar capacity to date is 5MW in the US, with 177MW installation under development.

Figure 8.11 Fresnel reflectors are not as efficient as parabolic mirrors in Nevada Solar One concentrated solar power plant, but may be cheaper; these are made by Ausra.

Source: Wikimedia Commons, from Ausra media department, with GFDL licence

Advantages

- readily available;
- cheaper to make;
- hybrid operation possible;
- very high space efficiency around solar noon.

Disadvantages

- recent development, only small projects operating.

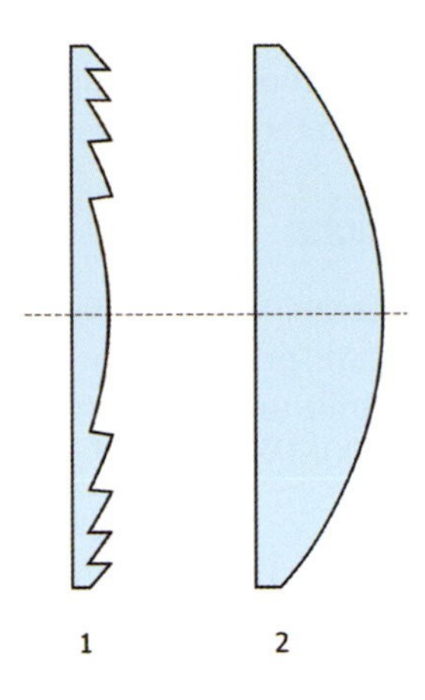

Figure 8.12 A cross-section of (1) a Fresnel lens and (2) a conventional plano-convex lens of equivalent power. The Fresnel lens is cut to approximate the shape of a convex lens but is much cheaper to produce.

Source: © Wikimedia Commons

Central receiver or solar tower

A circular matrix of heliostats (large mirrors that track the sun) focuses the sun's rays onto a central receiver on a tower. A heat-transfer substance (water/steam, molten salts or air) uses it to generate superheated steam for a turbine at temperatures around or over 1000°C (1830°F). If pressurized gas or air is used, it

Figure 8.13 This 11MW solar updraught tower in Sanlúcar la Mayor, Spain, uses 624 tracking mirrors, each with a surface of 120m^2 (1292ft^2). They reflect sunlight onto the top of a 115m (380ft) tower where it heats water into steam to drive a turbine to generate electricity. Opened in 2009, this PS10 plant is planned to be the first of several that will eventually generate over 300MW of power for the region by the year 2013. Once finished, it is projected that the solar plant will provide enough electricity to power over 180,000 homes in the surrounding areas and will prevent over 600,000 tonnes of CO_2 from being released into the environment annually.

Source: © Wikimedia Commons

can replace natural gas in a gas turbine, to take advantage of the high efficiency (60 per cent and more) of modern gas and steam combined cycles. Existing commercial plants are in the range of 11–100MWe, but much larger ones are in development.

Applications

Such systems are appropriate for grid-connected plants, producing high-temperature process heat. The highest single unit solar capacity to date is a 20MWe unit under construction. Total capacity is approximately 50MW with at least 100MW under development.

Advantages

- good mid-term prospects for high conversion efficiencies, with an operating temperature potential beyond 1000°C (1830°F) (565°C [1050°F] proven at 10MW scale);
- storage at high temperatures;
- hybrid operation possible;
- better suited for dry cooling (see later) than parabolic troughs and Fresnel;
- better options to use non-flat sites.

Disadvantages

- needs wider scale proof of concept in commercial operation.

Parabolic dish

A parabolic dish-shaped reflector concentrates sunlight onto a receiver at its focal point. The receiver contains a fluid or gas (air) and is heated to

Figure 8.14 Six dish Stirling Systems developed by Schlaich Bergermann und Partner of Stuttgart, Germany, in operation at the Plataforma Solar de Almeía in Spain. The three dishes in the foreground are second-generation systems, which produce 10kW of power from a Solo Kleinmotoren engine.

Source: Wikimedia Commons/Sandia National Laboratory

approximately 750°C (1380°F). This is used to generate electricity in a small piston or Stirling engine, or in a micro turbine integral with the receiver. The dishes usually track the sun along one axis – east–west. The system is modular and any number of dishes can be installed in a system from one upwards, allowing for later expansion.

Applications

Parabolic dishes can be used in stand-alone, small off-grid power systems or clustered to comprise larger, grid-connected dish parks. The highest single unit solar capacity to date is 100kWe. Proposals exist for 100MW and 500MW plants in Australia and the US.

Advantages

- very high conversion efficiencies – peak solar to net electric conversion over 30 per cent;
- modular;
- most effectively integrate thermal storage and large plant;
- operational experience of first demonstration projects;
- easily manufactured and mass-produced from available parts;
- no water requirements for cooling the cycle.

Disadvantages

- no large-scale commercial examples;
- projected cost goals of mass production still to be proven;
- lower delivery potential for grid integration;
- hybrid receivers are still an R&D goal.

Environmental benefits and impacts

Compared to fossil fuel power plants, concentrating solar power seems more expensive. However, when environmental and hidden costs are factored in and the levelized costs are compared, CSP is cheaper than nuclear power and 'clean' coal. It is quicker to build and commission and it does not have any polluting or toxic legacy. Also, its cost is falling while those of nuclear power and fossil fuels are rising.

It is estimated that each square metre (square yard) of concentrator surface can avoid 200–300kg (420–630lb) of CO_2 each year, depending on configuration. The plants can cover a large area of land and this may have an impact depending on the location. Desert areas with low biodiversity or scientific interest are the most suitable, but large volumes of water for cooling are required for parabolic troughs and linear Fresnel reflectors. Because they need cooling at the end of the steam turbine cycle, these CSP plants can have a high demand for water, which may be problematic in arid areas. Such sites should therefore be near the sea or rivers. The Andasol power plant's annual water needs are about the same as what would be needed for the cultivation of crops, such as wheat, on the same site (although it wouldn't be as the land is arid) – about 870,000m^3 (190 million gallons) of water per year, mainly used for cooling the steam circuit, that is from the vaporization of water in the cooling towers. The water comes mostly from ground water drawn from wells on the site. Dry cooling of thermal plants (with air) is possible, but is more costly and less effective than wet cooling; the technology requires higher investment and the electricity produced can be 5–10 per cent more expensive than with wet cooling. Compact Linear Fresnel Reflector (CLFR) technology has a much better power output to land requirement ratio and is faster and cheaper to construct than other CSP. Agricultural or environmentally unique land is the least suitable. Local environmental impact assessments are essential.

CSP has a similar impact to any conventional thermal power plant (gas, coal), minus the pollution associated with fossil fuels, but increased by the quantity of materials – mostly metals and concrete – used for the collectors. In terms of their energy payback or amortization period – the time a power plant needs to produce the energy required to manufacture and install the equipment – solar thermal power plants have a relatively short period of about five months, which is low compared to other forms of renewable energy.[1] Wind power's payback is 4–7 months but 2–5 years are needed for PV power plants. CSP plants require less surface area in terms of the amount of energy produced per square metre than PV and wind power. Based on the experience of parabolic trough plants in the Mojave Desert, US, which have been operating since the 1980s, installations could last up to 40 years. Most of the materials employed can be recycled and used again for further plants.

Where is CSP suitable?

As with PV power stations, CSP technology is best suited to the sun-belt areas of the world: southern Europe, northern Africa and the Middle East, parts of India, China, southern US, northern Chile and Australia, where there can also be peak power demands for cooling. Solar thermal power uses direct sunlight – 'beam radiation' or direct normal irradiation (DNI) – requiring at least 2000kWh of sunlight radiation per square metre annually.

The best sites get over 2800kWh/m^2 a year, for example steppes, bush, savannahs, semi-deserts and true deserts, ideally located within less than 40° of latitude north or south – such as the Near and Middle East, Iran, the Mediterranean and the desert plains of India, Pakistan, the former Soviet Union, China and Australia. Sites subject to clouds, fumes or dust in the atmosphere are inappropriate and would be better off considering thin-film PV power stations, which are more tolerant of these conditions. One of the most interesting projects in development is the Ordos project in the Inner Mongolia region of China, using parabolic trough technology. When completed in 2020, it will have a capacity of 1000MW.

All other factors being equal, plants in the Southwest US or Upper Egypt will supply electricity at a cost 20–30 per cent lower than in southern Spain or the North African coast because these areas receive 30 per cent more energy from direct sunlight (2600–2800 compared to 2000–2100kWh/m^2 a year). But the world's luckiest areas in this respect are the deserts of South Africa and Chile, which receive almost 3000kWh/m^2 a year. These countries would benefit the most from CSP. At the other end of the scale, the cost of CSP-generated electricity would be highest in France, Italy and Portugal.

At an optimum site, 1km² (0.386mi²) of land might generate as much as 100–130GWh of solar electricity a year, the same as that produced by a 50MW conventional coal- or gas-fired mid-load power plant. This is more power than is often needed in such regions, and so plans are under way for 'supergrids' to connect them to areas of high population density. The most advanced scheme is a European supergrid; Germany is already seriously considering importing solar electricity from North Africa and southern Europe.

Current installations

At the time of writing, there are about 20 installations operating throughout the world, generating 736.15MW. An additional 33 are under construction – all but four in Spain – with the potential to supply over 2000MW (2GW). A further 10GW have been announced in the US and 2GW in Spain, with another 7GW in other countries. Spain is the pioneering country in this technology, with three of the world's first plants, including the Andasol solar power station (100MW) and the PS20 (20MW) and PS10 (11MW) solar power tower (Figures 8.7, and 8.13 respectively). The US has the largest plants – the 354MW Solar Energy Generating Systems power plant – plus the Nevada Solar One (64MW), also pictured above (Figure 8.11). The Chinese are planning a 2000MW parabolic trough plant in Mongolia.

Hybrid plants are totally feasible; in fact one already exists. Ausra's 9MWt CLFR plant, completed in 2008, supplies solar-generated steam to is neighbour, Macquarie Generation's 2000MW coal-powered Liddell Power Station, near Muswellbrook, in the Hunter Valley, Australia. This displaces a portion of its coal consumption to produce the plant's electric power; a symbol of the transition to a post-fossil fuel age.

Figure 8.15 Several sun-tracking CFLRs focus sunlight onto the same horizontal pipes, making this power station highly efficient. Located in Kimberlina, southern California, the plant is being expanded to 177MW, and the start-up company that built it, Ausra, was acquired in 2010 by the huge French nuclear firm, Areva, which is expanding into this area. CLFR technology is predicted to be the market leader in solar thermal technology because it is cheaper and faster to construct, and its water and land requirements are half that of its competitors, for the same output.

Source: © Ausra

Economics

The development of the technologies discussed here is on a steep learning curve, and the factors that will reduce costs are improvements in the technology, mass production, economies of scale and improved operation. Adding more CSP systems to the grid can help keep the costs of electricity stable, and avoid drastic price rises as fuel scarcity and carbon costs take effect.

Experience in the US reveals costs of about 15 US cents/kWh at sites with very good solar radiation. CSP doesn't need long-term subsidies, like fossil or nuclear power, but simply an initial investment in the plant; over their lifecycle, 80 per cent of the costs are in construction and associated debt, and only 20 per cent from operation. So, once the plant has been paid for, over the next 20–30 years, only the operating costs remain, currently about 3 cents/kWh. The electricity generated is cheaper than any from the competition, comparable to long-written-off hydropower plants.

Prospects for growth

CSP is undergoing a huge growth of interest. According to the European Solar Thermal Electricity Association (ESTELA) and IEA SolarPACES, the technology is mature enough to grow exponentially in the world's sun-belt. They claim that CSP installations were providing just 436MW of the world's electricity generation at the end of 2008. But projects under construction at the time of writing will add at least another 1000MW by around 2011. Shams One, a 100MW CSP plant in the Emirate of Abu Dhabi, has recently been announced.

In Australia, there is a plan for a coastal hybrid power station consisting of 150MW of combined cycle gas turbine and 50MW of parabolic trough solar thermal, which will produce electricity, desalinated water and commercially saleable salt. In the US, projects adding up to a further 7000MW are under planning and development plus 10,000GW in Spain, which could all come online by 2017. Parabolic dishes are highly modular and most of them are currently not large-scale commercial grid-connected systems, but are used in small local projects. One of the largest is Sterling Energy Systems' 'Suncatcher', a 12m (38ft) diameter dish with a Stirling engine providing 25kW. However, one project in development would install 70,000 of these in the Californian desert to generate 1750MW.

The most exciting project, which would help to continue John Shuman's dream of powering the world with giant power stations in the desert, is Desertec, which is possibly Europe's most revolutionary and ambitious energy project. It aims to supply 15 per cent of Europe, the Middle East and North Africa's electricity needs by 2050, by generating it around the coasts of the Middle East and North Africa and linking it to a high voltage direct current grid that will also be connected to north European offshore wind farms. Electricity will come from concentrating solar thermal power. Desertec is led by German companies, including E.ON, RWE and Siemens, but includes five companies from Spain, Italy, France, Morocco and Tunisia. Paul van Son, chief

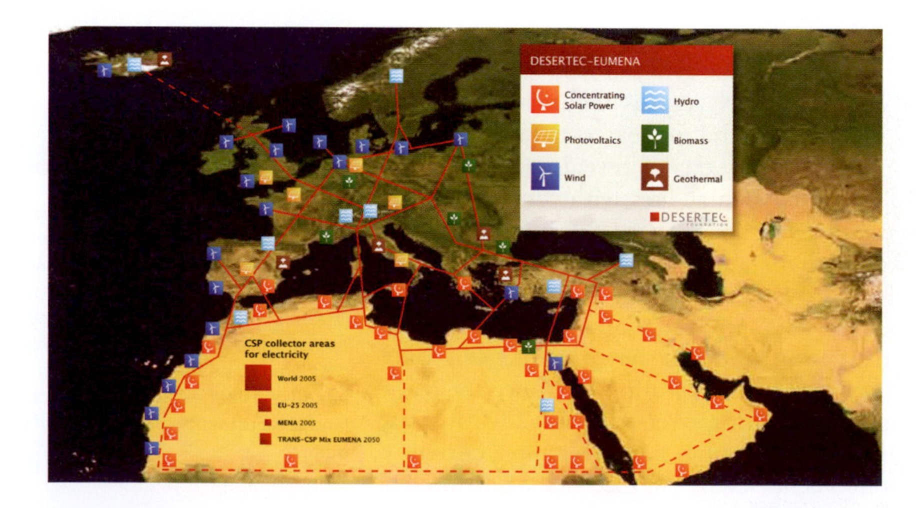

Figure 8.16 A map of the direct current high voltage electricity grid proposed by Desertec, to link CSP plants in the sun-belt with other renewable plants around Europe. Dr Thiemo Gropp, Director of the Desertec Foundation, believes that despite short-term delays, cooperation with democratic regimes in the area will help in the longer term.

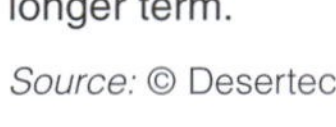
Source: © Desertec

executive of the Desertec Industrial Initiative, said the €400 billion long-term plan would start delivering its first projects within a decade. One of the initial jobs is to build the grid connectors.

Desertec's backers hope that it will also provide North Africa and the Middle East with desalinated water. If saltwater from nearby coasts is used instead of drinking water for the cooling units, a 250MW collector field may be used to operate a 200MW turbine and 100,000m^3 of drinking water may be produced a day (over four million litres per hour) through the process of water desalination.

According to the Global CSP Outlook 2009, under an advanced industry development scenario, with high levels of energy efficiency, CSP could meet up to 7 per cent of the world's projected power needs in 2030 and a full quarter by 2050, employing as many as 2 million people. Even with a set of moderate assumptions for future market development, the world would have a combined solar power capacity of over 830GW by 2050, with annual deployments of 41GW. This would represent 3.0–3.6 per cent of global demand in 2030 and 8.5–11.8 per cent in 2050.

Solar towers

A 'solar tower' (also known as a solar chimney) is a third type of electricity generating technology that exploits solar energy. This one uses air pressure differences to drive a turbine. A large, circular, ground-level greenhouse, exposed to sunlight, draws air in and heats it up. The hot air expands and is sucked up the chimney by a powerful stack effect, where it turns turbine blades that rotate coils in generators to produce electricity.

There are no current large-scale examples. A 50kW prototype, designed by engineer Professor Jörg Schlaich, was constructed in Manzanares, Spain, and operated for eight years before being decommissioned in 1989. A company called EnviroMission is hoping to build a 200MW power station in Arizona, US. It would contain 326.25MW turbines and the tower would be 1km (3280ft) high. The base area would cover several square kilometres. As the air rises, it

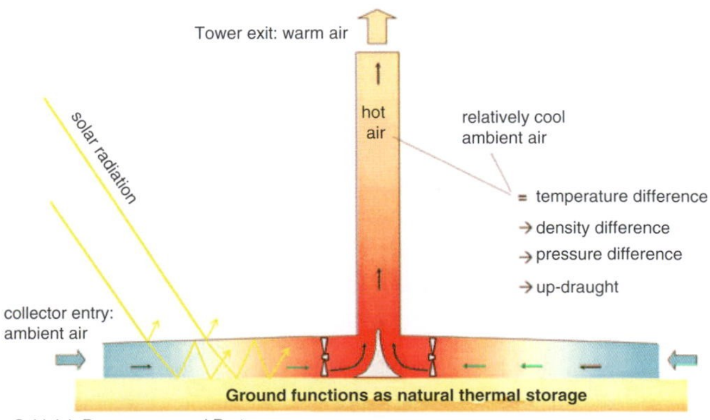

Figure 8.17 EnviroMission's pilot solar tower and a conceptual diagram.

Source: Enviromission

is expected to accelerate to 70km/hr (43mph). As it consumes little water, EnviroMission thinks this technology is perfect for dry, hot areas of Australia.

Another – untested – version of a solar tower heats water at ground level, using the steam to generate electricity. It could theoretically operate round the clock as water is an efficient heat storage medium for night-time use, and a refinement of the principle uses a patented solar storage capacity for certain salts. The idea, proposed by David Daudrich, a German scientist, is to situate such generators near the coast in a desert area in order to also produce a supply of clean desalinated water, as well as rain clouds that might bring precipitation to the area around the power station. At the time of writing, the development companies involved are still seeking finance for these large-scale projects.

Thermoelectrics

At the other end of the scale of solar electricity-generating technologies are thermoelectric, thermionic or thermovoltaic devices. These convert a temperature difference between dissimilar materials into an electric current. The effect was discovered in 1821 by the physicist Thomas Seebeck and explored by Jean Peltier. It is sometimes called the Peltier-Seebeck effect. First proposed as a way of storing solar energy by Augustin Mouchot in the 1880s, it is now one of several cutting-edge nanotechnologies in development around the world. The Peltier effect can also be used for cooling.

Some materials, such as molecule-thin diamonds, discharge electrons when heated. In a thermoelectric cell, these jump across a vacuum (to avoid convection), a few hundred micrometres wide. They are collected by a thin film on the opposite side, and thus generate an electric current. Physicist Neil Fox at the Centre for Nanoscience and Quantum Information in the University of Bristol, UK, is confident that since they use no moving parts they could become very reliable and efficient. An American company, MicroPower Global, owns a similar patent and claims to be about to roll out modules of microchips utilizing the principle. A thermoelectric application would consist of a thermoelectric chip, a module, a sub-assembly and the encompassing system.

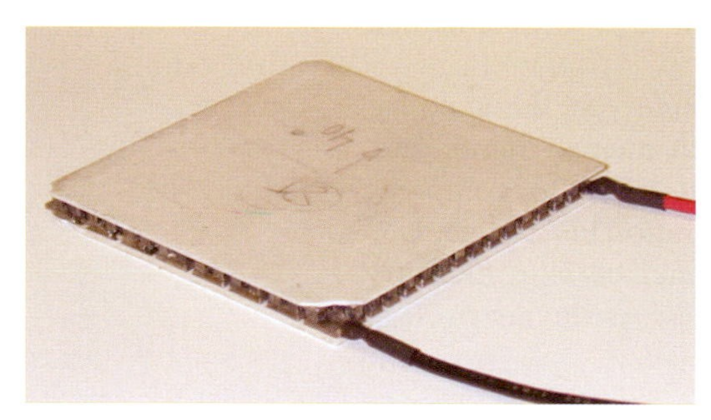

Figure 8.18 A Peltier element, which works by passing a current through differing metallic conductors to produce a positive or negative heat difference. The process can be reversed.

Source: © Creative Commons

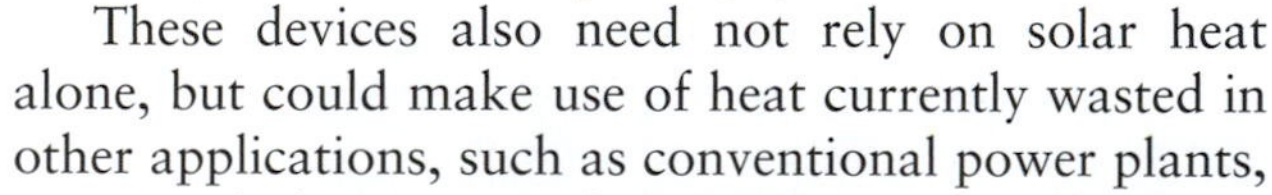

These devices also need not rely on solar heat alone, but could make use of heat currently wasted in other applications, such as conventional power plants, internal combustion engines and data server farms. These typically waste 60–70 per cent of their energy as heat. Efficiencies of thermoelectric devices in the laboratory are currently in the order of 10 per cent. There are many technical hurdles to overcome before they reach commercial viability.

Note

1 According to lifecycle assessments of components and land-surface impacts conducted by the European Solar Thermal Electricity Association (ESTELA) and IEA SolarPACES.

9 Conclusion

In the different chapters of this book, we have seen various forecasts for the growth of different technologies and practices using solar energy. The main areas of growth are undoubtedly to be found in solar building, solar cooling, solar power stations, off-grid small-scale rural electrification and distributed energy.

The use of passive solar architecture for heating and cooling of buildings has huge potential, as cities grow and more and more of the world's human population crowds into them. The utilization of solar gain, and standards similar to the 'Passivhaus' standard, can help to curb the rate of growth in demand for electricity for these purposes. The government-level policy change that will most affect this is modification to national building regulations, and, consequently, a massive rollout of training for building control officers so they understand the principles of passive solar, superinsulated buildings and how they should be applied.

Solar cooling, which is now tiny in terms of its market penetration, has a long way to go, but huge potential, whether it is passive (through building design) or active (through solar-powered chillers). It is a blindingly obvious fact that seems to have largely escaped widespread notice, that solar power is available for cooling purposes at precisely the time when it is most needed.

There is bright hope for CSP, using proven technology that is over 130 years old, to generate power in hot, open-skied areas of the world. These power stations could be built relatively quickly, certainly more so than nuclear power stations, although the grid infrastructure also needs to be constructed to deliver the power to where it is most needed. Solar power stations have the following major advantages over nuclear power:

- the fuel source will never run out;
- therefore the technology will never become obsolete;
- it is predictable and reliable (in sun-belt areas);
- it is completely free;
- it is much safer;
- it is cheaper, when calculated on a levelized cost basis per megawatt generated;
- its mining does not cause conflict (as extraction does in several parts of Africa);
- there is absolutely no pollution, radiation danger to workers or toxic legacy for tens of thousands of years.

The most promising of the CSP technologies in terms of speed of construction, water-efficiency and cost-efficiency is CFLR. We can expect to see a great many of these plants built in the coming years in sun-belt areas.

For the billions of people who still have no access to electricity, and the millions who have no access to fresh, clean water, solar power, along with wind energy (depending on location), has the potential to transform lives and help reach many of the United Nations Millennium Development Goals.

Distributed energy, which is a name for households, buildings or communities generating their own power locally for electricity or heating/cooling, is another huge growth area. The benefits of distributed energy are: local ownership, reduced transmission losses and greater efficiency. At too small a scale, however, the economy of scale is lost because of unnecessary duplication of system components such as inverters (house-by-house, instead of street-by-street). Renewable energy at village, street or town scale therefore has a part to play in community regeneration, as communities come together to take responsibility for their own energy consumption and learn what it means.

Energy delivery has been taken for granted for too long. Consumers of energy have a lot to learn about how their pattern of use affects overall demand, and how they can reduce their energy consumption. It is also hoped that the rollout of the 'smart grid' can help to manage or smooth out peak loads and reduce overall requirements for generation capacity.

Figure 9.1 CESA-1 solar tower power station, CIEMAT-PSA, Almería, Spain.

Source: © SolarPACES

Box 9.1 General conversion efficiencies of different conversion systems

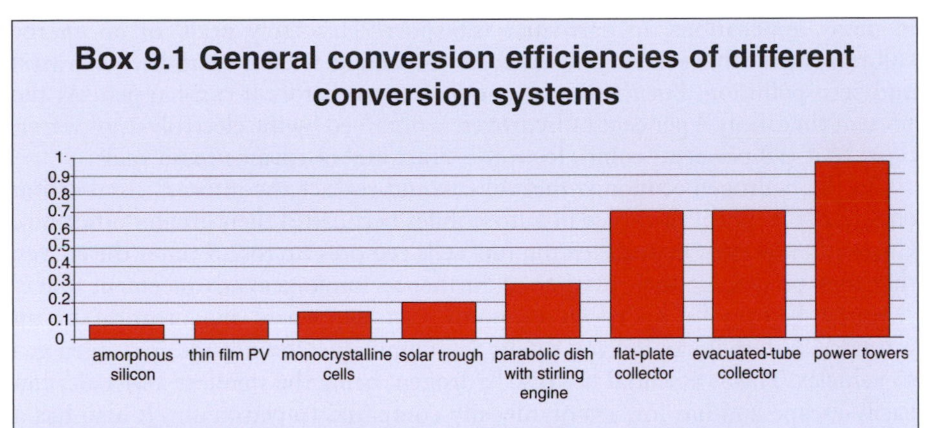

Figure 9.2 General conversion efficiencies of different conversion systems.

Table 9.1 Conversion efficiencies.

Technology	approximate efficiency	output
Amorphous silicon	0.08	electricity
Thin-film PV	0.1	electricity
Monocrystalline cells	0.15	electricity
Solar trough	0.2	electricity/heat
Parabolic dish with Stirling engine	0.3	electricity
Flat-plate collector	0.7	heat
Evacuated-tube collector	0.7	heat
Power towers	0.97	electricity/heat

Figure 9.2 and Table 9.1 give a rough idea of the relative conversion efficiencies of different technologies. Gross conversion efficiencies are determined by net generating capacity over the solar energy that falls on the total area of the solar plant (Note: solar dishes or troughs occupy only a fraction of the total area of the power plant). Much depends on the local conditions. And, at the end of the day, what is really important is the cost per kWh produced.

In terms of making optimum use of surface area, a single solar dish Stirling engine installed at Sandia National Laboratories' National Solar Thermal Test Facility produces as much as 25kW of electricity, while its footprint is 100 times smaller than Spain's solar updraught tower. A 500MW Southern California Edison (SCE) and Stirling Energy Systems 1800ha (4500ac) solar power-generating station in Southern California is expected to extract about 2.75 per cent of the insolation (1kW/m^2) that falls on its 18.2km^2 (7mi^2). The 49.9MW AndaSol commercial parabolic trough solar thermal power plants, near Guadix in Andalucía, Spain (total area = 195ha), has an annual average gross conversion efficiency of 15 per cent (the collectors themselves have 50 per cent; the rest is system losses).

The hydrogen economy

There has been much talk of a 'hydrogen economy' in the last 40 years. The dream has been that hydrogen gas, generated by solar power from the electrolytic splitting of the water molecule, could be used eventually to replace fossil fuels in many applications, in particular transport. This ‘holy grail’ of an energy solution would have low running costs (assuming wide availability of water) and zero pollution. But there is a long way to go before it can happen. At the present time, only 4 per cent of hydrogen is obtained by the electrolysis of water; most of it (90 per cent) comes from the ‘cracking’ of various fossil fuels.

In the hydrogen economy, fuel cells would replace the internal combustion engine as the motor of choice in automobiles because of their greater efficiency. Currently however, manufacturing fuel cells requires up to 2.5 times the energy that they produce in their lifetime. A further technological advancement that is required before the hydrogen economy can happen is the creation of an infrastructure that can deliver, safely, the flammable compressed hydrogen gas to vehicles. This is essential because hydrogen, being the smallest molecule, can easily escape and has low energy density compared to petroleum. It also has a tendency to render brittle pipes and tanks containing it, unless they are specially coated. Many different types of storage mechanism have been tried out, including underground storage – successfully employed by ICI for many years.

To produce hydrogen using solar electricity requires up to 50kWh of electricity per kilogram of hydrogen produced (although this figure is reducing, see below). High temperatures – 800–1200°C (1472–2192°F) – are required to split water into its two constituent elements. Concentrating solar power is being tested at an industrial scale to achieve this, at the Hydrosol-2 100kW pilot plant at the Plataforma Solar de Almería in Spain, which has been operating since 2008 (Figure 9.1). As with all solar power, it is modular and, if successful, can be readily scaled up. The projected costs of hydrogen produced by CSP and electrolysis, assuming solar thermal electricity costs of $0.08/kWhel, range from $0.15–0.20/kWh, or $6–8/kg H_2.[1] The US target for 2017 is $3/gge (gasoline gallon equivalent; 1 gge is about 1kg H_2), and the EU target for 2020 is €3.50/kg H_2. Besides electrolysis, thermo-chemical production of hydrogen using concentrated solar power is also being actively investigated. American and European pilot studies indicate that this can be cost-competitive compared with the electrolysis of water using CSP. It can even become competitive with conventional fossil fuel-based processes for generating hydrogen at current fuel prices, especially if credits for carbon dioxide mitigation and pollution avoidance are applied. Hydrogen generated centrally like this would still have to be stored, or piped to the point of use.

But what about electrolytically generating the hydrogen at the point of use using PV? Such a system is being pioneered by Honda in the form of a home or workplace refuelling appliance for fuel cell electric vehicles (Figure 9.3). The prototype concurrently produces enough hydrogen (0.5kg) in eight hours for daily commuting (16,093 km per year, 290 km per day; 10,000 miles per year, 180 miles per day). It employs a patent 48-panel 6kW PV array using thin film CIGS, that is 16kWh/kg of hydrogen. It uses an innovative high differential pressure electrolyte, which eliminates the need for a compressor and is highly compact. It is designed to work in tandem with the company’s FCX Clarity vehicle and a ‘smart grid’ energy system that can also export surplus energy to the grid.

Figure 9.3 Honda's PV-powered home or workplace fuelling system for hydrogen-powered fuel cell vehicles.

Source: © Honda

Only some car companies are researching hydrogen as a fuel. This is because it is less efficient, currently, than alternatives. Electric vehicles running on, say, a lithium-ion battery, are around three to five times more efficient. It is thus likely that at least in the near or mid-term future, we will see increasing numbers of electric vehicles on the roads, as more and more renewable or low-carbon electricity from many different sources is introduced into countries' national grids.

Breaking the fossil-fuel addiction

The Earth's history has provided us with a great gift of fossil fuels. Over the last 300 years, human beings have used it in increasingly ingenious ways to replace human- and animal-power, which were the chief means of production and transport in previous centuries. This has produced the extraordinary achievements and civilization whose benefits about one-fifth of the population of the world now fully enjoy, and which the rest wish to. Before these fuels were discovered and the Industrial Revolution began, energy was also supplied by the renewable sources of water and wind power as well as burning biomass, and these have a continuing part to play.

But as long ago as the 1870s, visionary solar pioneers such as Augustin Mouchot foresaw the time when coal would run out and they began to develop alternatives that could deliver the same benefits from solar power. In 1913, Frank Shuman, who designed the world's first solar power station, dreamt of a completely solar-powered world. It was theoretically possible then, as indeed it is now. What has prevented its breakthrough has been the easier option of fossil fuels. Yet, the competition between nation states for access to these resources has time and again over the last century brought violent conflict, suffering, widespread destruction and loss of life. The presence of oil, coal and gas in a territory has been a curse as much as a blessing. With growing awareness of the impact of climate change, their aspect as curse on the global scale has become increasingly apparent.

Nowadays, the phrase 'energy security' is being used by those who want to see local, sustainable sources of clean energy replace dirty fossil fuels. This is because the sun, wind and other renewable sources of energy are available abundantly, everywhere on the planet, with no need for conflict over their use. Looking at the history of solar power, it is clearly obvious that its development has suffered as a result of the abundance of fossil fuels. The world's economy is currently predicated upon their use. Despite all the scientific evidence of the imminence of catastrophic climate change as a result of our continued use of these fuels, the companies and economies which rely on them are as enthusiastic as ever to exploit them. This is occurring in more and more technically and geographically challenging locations, from the seabeds off the Brazilian coast to those beneath the Arctic Ocean, ironically now more available precisely because of global warming.

In the Greek myth of Prometheus, this human stole fire from the gods for the benefit of all humanity. But the gods punished humanity by giving them Pandora's box, which, when opened, gave them for the first time, sickness, worry and death. One cannot help but recognize parallels with fossil fuels and nuclear power. Humanity – or its leaders – are now faced with a clear choice: whether to stick with the status quo and vested interests that aggressively promote as inevitable a continued dependence on fossil fuels; or whether to accelerate the deployment, research and development into solar and other renewable, sustainable technologies and practices. The potential of these technologies is completely clear and proven. The scientific case for a runaway greenhouse effect, if things continue as they are, has been conclusively established. The stakes could not be higher, nor the choice more stark.

Note

1 *Solar Fuels from Concentrated Sunlight*, SolarPACEs/IEA, 2009.

10
Resources

How much energy is available?

To use solar energy, it is necessary to know how much is available in a particular location. Most countries now provide insolation maps which illustrate the amount of solar energy available over the year in a given location. Figures 10.4 and 10.5 show one for Africa and one for Europe, but they show different information. Figure 10.4 shows the annual average of each day's sunshine falling on a horizontal surface taken over the years given. The key is in watt-hours per square metre. Figure 10.5 shows the figure in kilowatt-hours per square metre for optimally inclined south-facing PV modules. But it also includes an estimate of the amount of electricity that could be generated by a 1kW system with an efficiency of 75 per cent – although this might vary in practice.

MESoR

The EC-funded project MESoR (Management and Exploitation of Solar Resource Knowledge) is a comprehensive portal giving user-friendly access to several free sources of solar energy data using a map-based graphical user interface (GUI). The sources include NASA, PVGIS, SoDa, HelioClim, NCEP, SWERA, METEONORM and MeteoTest, plus others. Users input their locations and the output format can be chosen. Map layers can be shown in Google Earth. Work is on-going to combine and harmonize the various data sources and provide quality control to improve the reliability of the final figures. Some of these data sources are discussed below. Higher resolution maps are becoming available to improve the precise siting of PV power stations.

HelioClim-3

HelioClim-3 is a service providing irradiance data from 15 minutes to a month, for Europe and Africa, computed from images taken by the Meteosat satellites. Meteonorm shows monthly averages (10–20 years) of global radiation resulting from interpolation of measurements made in meteorological networks. Solar Energy Mining (SOLEMI) is a service also providing irradiance data derived from Meteosat satellite images, including global horizontal (GHI), and direct at normal incidence (DNI). The dataset is from 1991–2005 (Europe and Africa), 1999–2006 (Asia). SoDa is similarly derived global irradiation data with free maps.

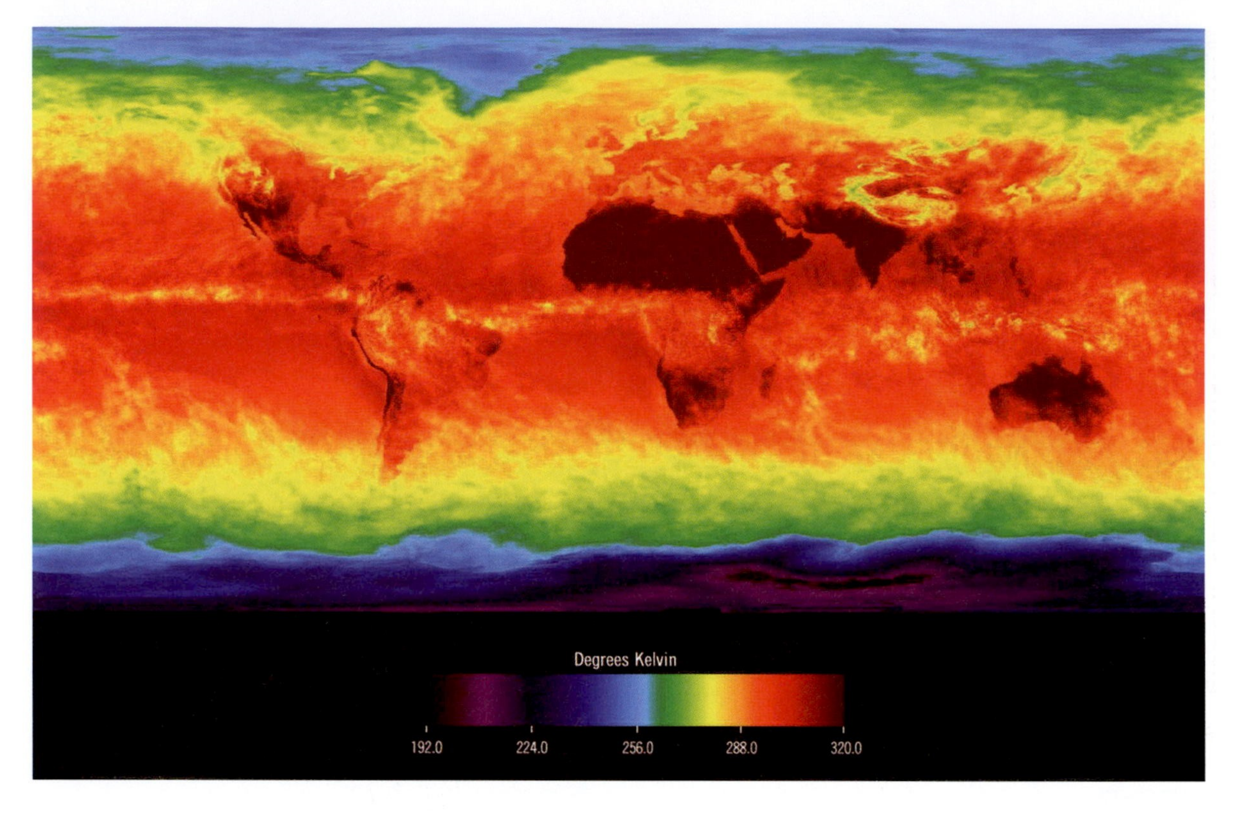

Figure 10.1 A NASA graphic illustrating the sunniest (dark red) spots in the world.

Table 10.1 Summary of providers of solar information.

Product	Area	Period	Provider	Temp. resolution	Spatial resolution	Access	Price
NASA	World	1983–2005	NASA	average daily profile	100km	eosweb.larc.nasa.gov/sse	free
METEONORM	World	1981–2000	Meteotest	synthetic hourly/min	1km (+SRTM)	CD or www.meteonorm.com	€410
SOLEMI	World	1991–>	DLR	1h	1km	on request	on request
HelioClim	World	1985–>	MINES-ParisTech	15min/30min	30km/3–7km	www.helioclim.org	on request
EnMetSOL	World	1995–>	Univ. of Oldenburg	15min/1h	3–7km/1–3km	on request	on request
(Satel-Light)	Europe	1996–2001	ENTPE	30min	5–7km	www.satel-light.com	free
PVGIS	Europe	1981–1990	JRC	average daily profile	1km (+ SRTM)	re.jrc.ec.europa.eu/pvgis	free
ESRA	Europe	1981–1990	MINES-ParisTech	average daily profile	10km	CD	€380

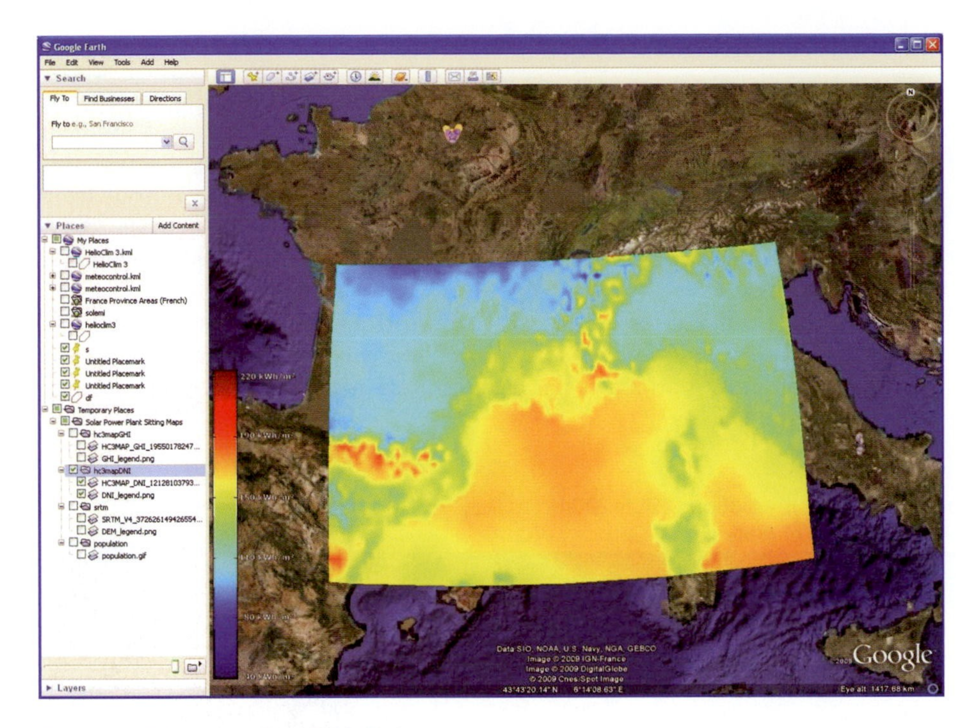

Figure 10.2 Screen grab of MESoR data.

Source: © EC/MESoR

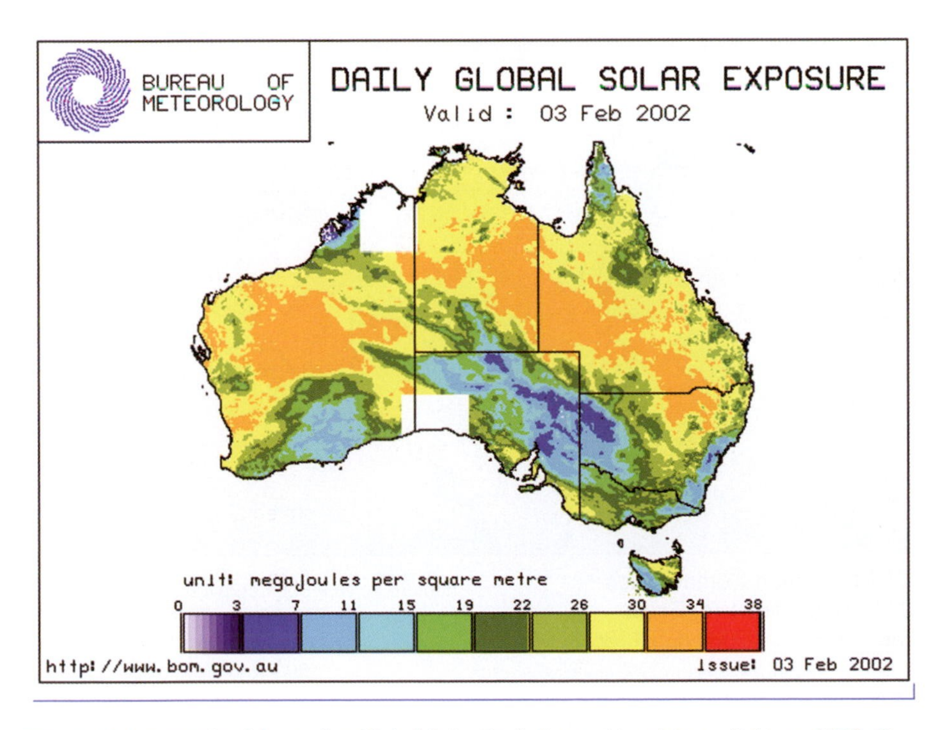

Figure 10.3 Example of Australian Global Solar Radiation archive data available on MESoR.

Source: © EC/MESoR

PVGIS and PVWatts

PVGIS (at http://re.jrc.ec.europa.eu/pvgis) contains the following resources: solar radiation data for Europe, Africa and South-West Asia, and ambient temperature for Europe, plus terrain and land cover. It can assess the available solar radiation for fixed and sun-tracking surfaces and yield the output from grid-connected PV and the performance of standalone PV (Africa only) dependent upon the slope, azimuth (angle of the sun), installed peak power and tracking options.

In the US, similar information is available from the National Renewable Energy Laboratory as PVWatts software on the website www.nrel.gov/rredc/pvwatts. They show national solar PV resource potential and CSP resource potential for the US.

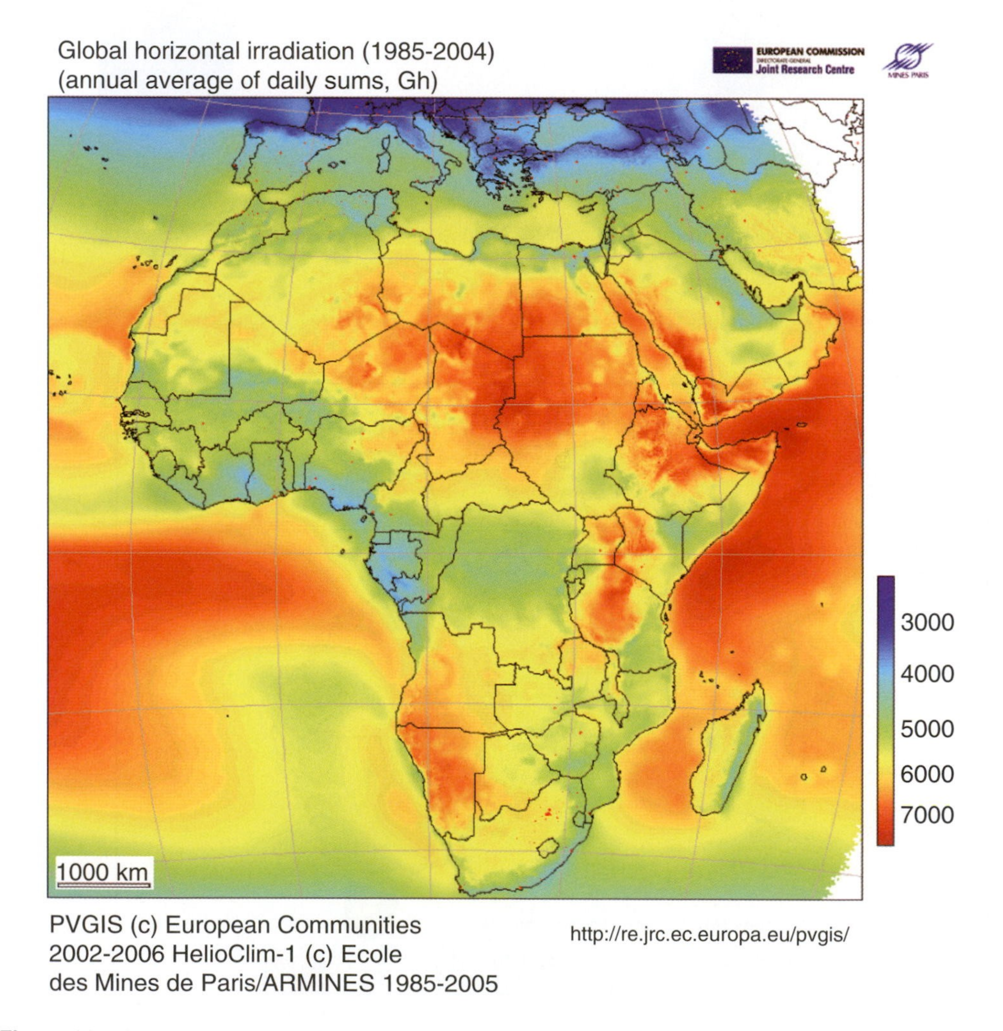

Figure 10.4 A sample map of solar insolation in Africa taken from the PVGIS European Union website.

Source: PVGIS © European Communities, 2001–2008

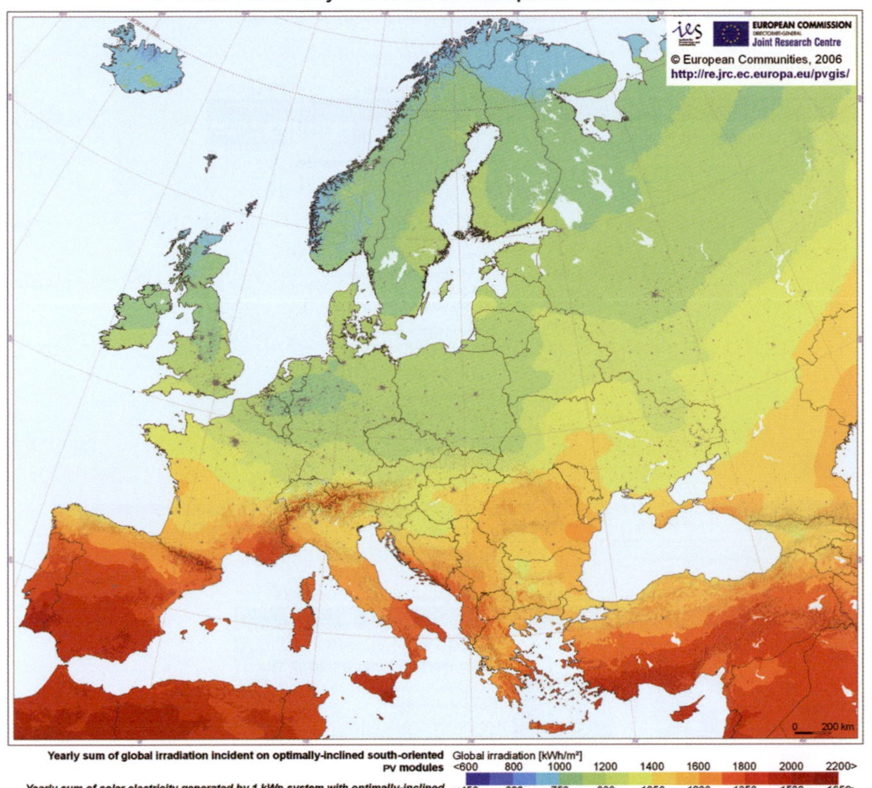

Figure 10.5 A sample map of solar insolation in Europe taken from the PVGIS European Union website.

Source: PVGIS © European Communities, 2001–2008

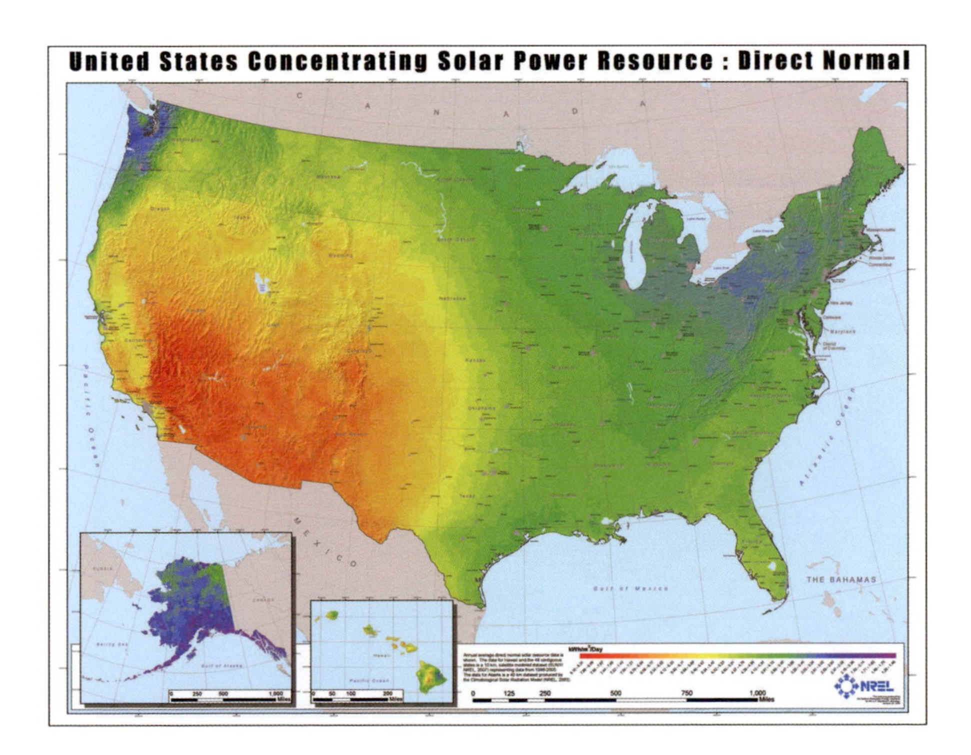

Figure 10.6 Insolation map for the US showing annual average direct solar resource data in kWh/m², available for concentrated solar power purposes. The darkest red indicates a maximum of 8.31kWh/m² per annum, while the minimum (lilac) is 1.25kWh/m² per annum.

Source: National Renewable Energy Laboratory, 2008

Figure 10.7 Other screenshots displaying the capabilities of PVGIS to respond to user-inputted data.

Source: PVGIS

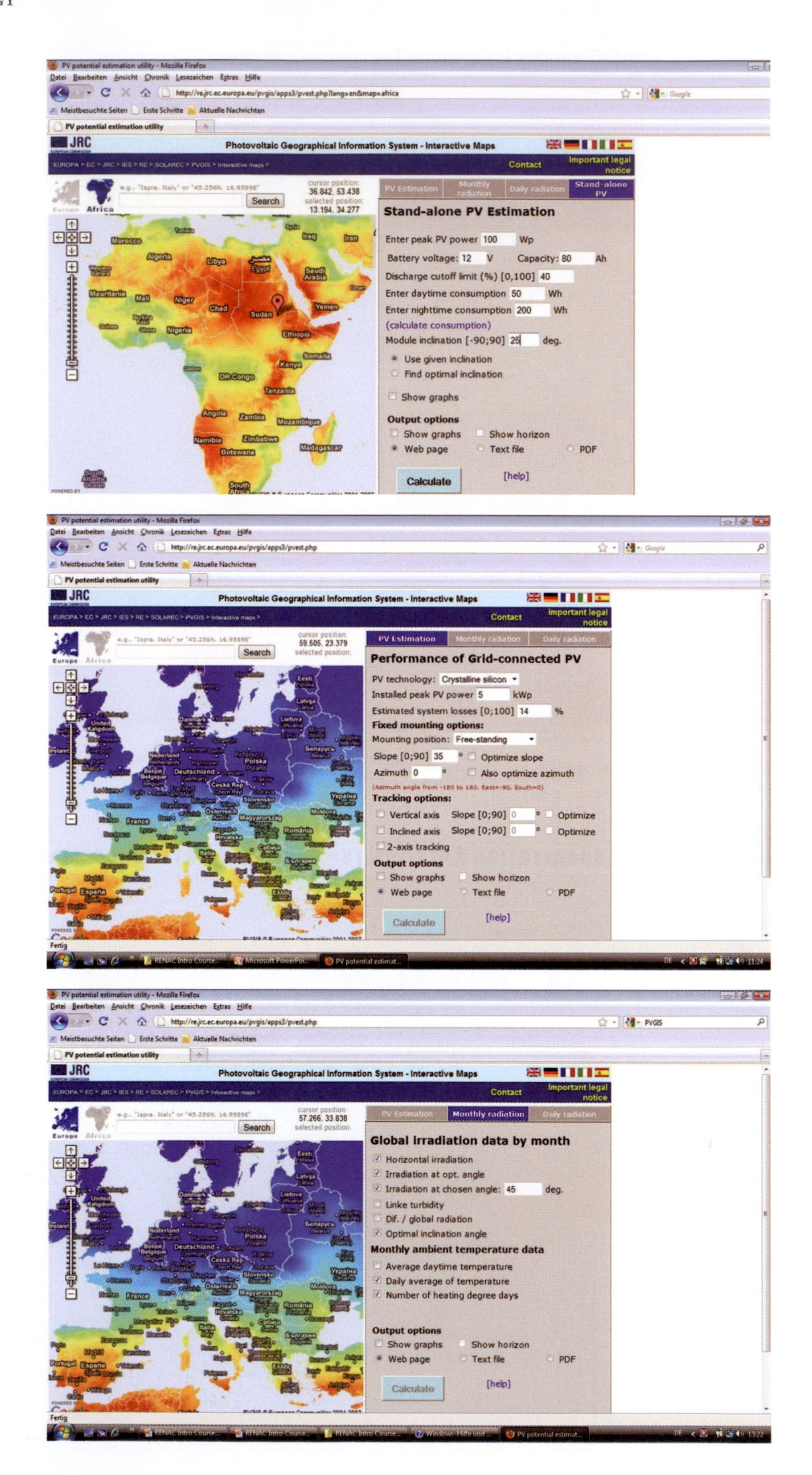

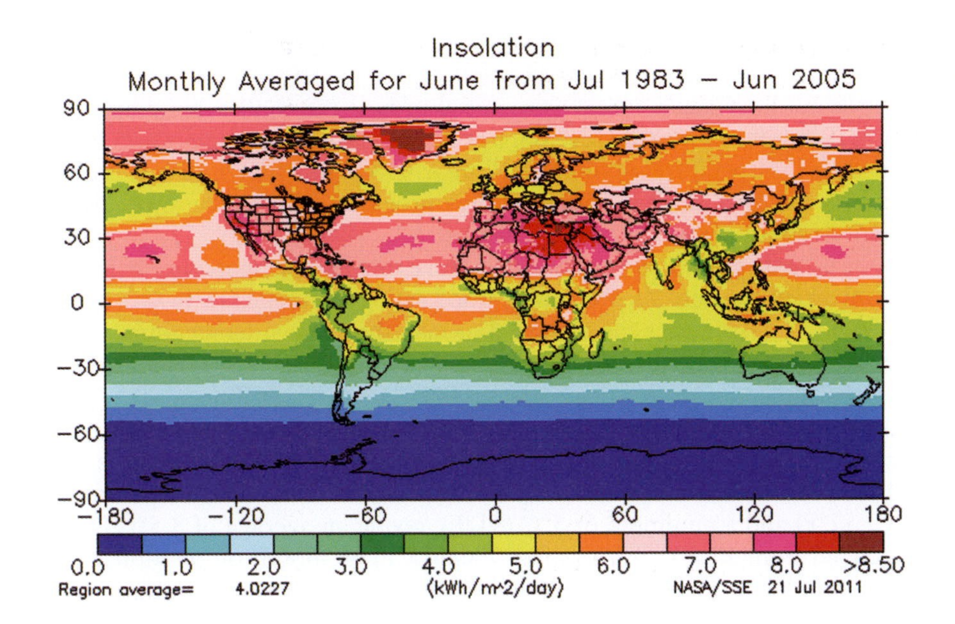

Figure 10.8 NASA has an interactive database that is freely available where you can choose a month and a location in the world and see, for example, how much cloud cover there is on average.

Source: PVGIS, NASA: www.eso.org/gen-fac/pubs/astclim/espas/world/ION/ion-cloud.html

NASA's Surface Meteorology & Solar Energy (SSE) Database

NASA's SSE is an interactive database of solar insolation (as well as rain and wind) information for regions of the world. It contains over 200 solar energy and meteorology parameters averaged from 10 years of data. Over 1,000,000 maps a month are generated from user requests. Users choose what information they want at the bottom of the page, and then select the time of year and the region of the world they want the information for. It is available at: http://eosweb.larc.nasa.gov/sse.

METEONORM

METEONORM refers to itself as a Global Meteorological Database for Engineers, Planners and Educators. It is a comprehensive meteorological reference, incorporating a catalogue of meteorological data and calculation procedures for solar applications and system design at any desired location in the world. Based on over 25 years of experience in the development of meteorological databases for energy applications, it has 28 different output formats as well as a user-definable output format. It is available for purchase from Meteotest at www.meteonorm.com.

TRY

Test reference year (TRY) files provide typical annual profiles of exterior climate data, such as ambient temperatures, wind direction and velocity, precipitations as well as direct and diffuse irradiances. An excellent free source of TRYs is the US Department of Energy's site at www.eere.energy.gov/buildings/energyplus. The site provides hourly climate data for many locations worldwide in the .epw format.

Degree days

To work out how much energy is required to heat (or cool) a building, it can be useful to refer to heating (or cooling) degree days. Degree days are derived from daily temperature observations in a given location and the heating requirements for a specific structure. They provide a rough way to predict the amount of energy required over a given period.

A heating degree day is worked out relative to a base temperature; in the UK, the convention is to use 15.5°C (60°F) (the heating requirement). In Germany, it is 17°C (63°F). To find it for a given day, take the average temperature on any given day and subtract it from the base temperature. If the result is above zero, that is the number of heating degree days (HDDs) on that day. If the result is below or equal to zero, it is ignored. These numbers for each day are added up over a year to find the total number of degrees of heating required for all the days.

For example, a location on one day might have a maximum temperature of 14°C (57°F) and a minimum of 5°C (41°F), giving an average of 8.5°C (47°F). Subtracted from 15.5°C (60°F) gives 7°C (45°F). A month of 30 similar days might accumulate 7 × 30 = 210. A year (including summer temperatures above 15.5°C (60°F)) might add up to 2000. The rate at which heat needs to be provided to this hypothetical dwelling is the rate at which it is being lost to the outside. This rate, for 1°C (1.8°F) temperature difference, is simply the U-value of the dwelling (as calculated by averaging the sum of the U-values of all elements) multiplied by the area of the dwelling's external surface.

Multiplying the rate at which a building is losing heat by the time (in hours) over which it is losing heat reveals the amount of heat lost in Wh (or Btu): this is exactly the amount of heat that needs to be provided by the heating sources. (To convert Wh to kWh, divide by 1000.) Therefore if the external area of the dwelling (walls, roof and floor) is A, its average U-value is U and the number of degree days is D then the amount of heat required in kWh to cover the period in question is:

$$A \times U \times D \times 24 / 1000$$

where the multiplier 24 is needed to get the value in kWh rather than kW-days or simply Btu in the case of US units. The same method is used (without dividing by 1000) for US units with the result being in Btu rather than in kWh.

For example, if the building surface area is 500m^2, the average U-value is 3.5 and the degree days number 2000 then the annual amount of heat required is:

$$500 \times 3.5 \times 2000 \times 24 / 1000 = 84{,}000\text{kWh}$$

In the course of a heating season, the number of HDDs for New York City is 5050 whereas for Barrow, Alaska, is it 19,990. Therefore, the same house will take four times as much energy to heat in Alaska as in New York. Meanwhile, in Los Angeles, which has an HDD of 2020, you would need only 40 per cent of the energy required in New York. The same method can be used for calculating the amount of energy required to cool a building, for the days in

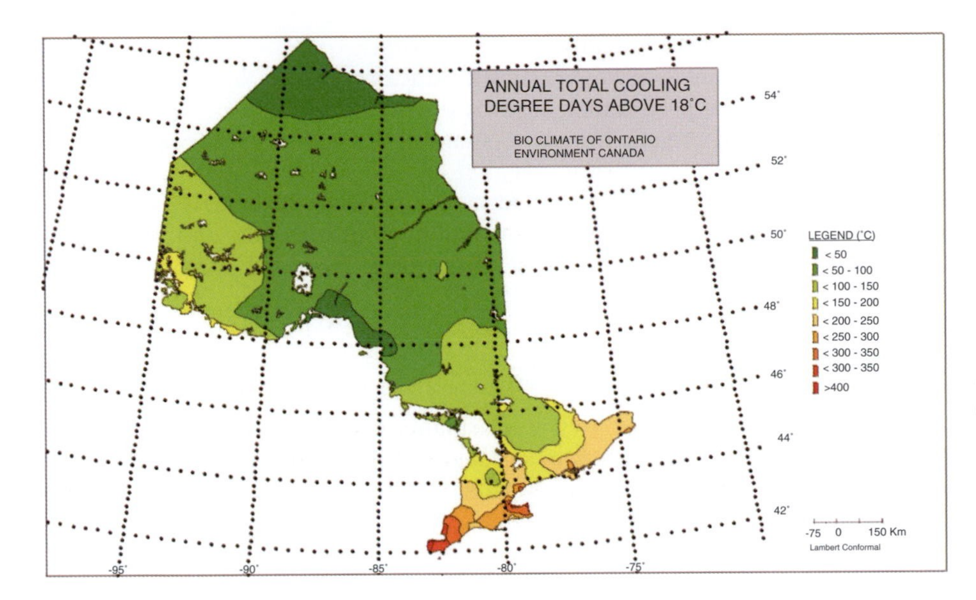

Figure 10.9 Cooling degree days – map of cooling degree days in Canada above 18°C (64°F).

Source: From the Canadian Integrated Mapping and Assessment Project at www.can-imap.ca

which the outside temperature is greater than the preferred inside temperature. This method is not foolproof, since there are other factors involved, such as the heat given off by appliances and people, wind factors, exterior shading, and so on. Also, heat requirements are not linear with temperature and superinsulated buildings can have a lower target point.

Note: Take care when using degree days to compare climates in different countries, because they may have different baseline temperatures, and because of the use of the Fahrenheit scale in the US and the Celsius scale almost everywhere else.

To convert °F HDD to °C HDD: °C HDD = (5/9) × (°F HDD)
To convert °C HDD to °F HDD: °F HDD = (9/5) × (°C HDD)

Because HDDs are relative to a base temperature, you don't need to add or subtract 32 when converting between Celsius and Fahrenheit.

Degree days are available worldwide for cooling and heating calculations at www.degreedays.net. The US data is available at the Climate Predication Center at www.cpc.noaa.gov/products/analysis_monitoring/cdus/degree_days. Software is often available online to calculate cooling and heating degree days and energy performance, using regional climatic conditions available from local weather services.

Other web resources

Case studies and technical information

The most valuable online resources, for case studies, good technical information and innovation, are provided by IEA's special projects. Principally these include:

- solar heating and cooling at www.iea-shc.org;
- district heating and cooling at www.iea-dhc.org;
- PV power system programme at www.iea-pvps.org;
- solar power and chemical energy systems at www.solarpaces.org.

There was also an excellent European Union project, SOLARGE, which promoted solar thermal technology on buildings of all types and finished in 2007, but its website still contains useful resources for good practice at www.solarge.org.

Real-time PV system examples

A website called 'Solarplanez' (www.solarplanez.com) allows anyone to search for examples of PV installations in selected countries, and view or download their location, size and real performance data, in many cases in real time. Here is an example:

Location: Switzerland – Genève
Connection: September 2008
Altitude: 414m
Power rating: 3120Wc
Production expected: 3000kWh/yr
Actual production (2009): 3297kWh/yr
Inverter: Sputnik – SOLARMAX SM3000S
Area: $24m^2$
modules: 24 × Solar Fabrik – SF 130/4-130
Angle: 15° (optimum: 34°) (source: PVGIS)
Orientation: 20°/South (optimum: 1°) (source: PVGIS)
Annual loss compared to optimum: 4.35% (source: PVGIS)
Installer: Solstis.

Printed resources

Books

Martin, C. L. and Goswami, D. Y. (2010) *Solar Energy Pocket Reference*, International Solar Energy Society, Earthscan, London

Haggard, K., Bainbridge, D.A. and Aljilani, R. (2010) *Passive Solar Architecture Pocket Reference*, Earthscan, London

Architectural Graphic Standards, 10th edn., available from John Wiley & Sons, Inc. Publishers. A source of information on sun angles and solar path diagrams for calculating shading and windows.

Hankins, M. (2010) *Stand-alone Solar Electric Systems: The Earthscan Expert Handbook for Planning, Design and Installation*, Earthscan, London

Antony, F., Durschner, C., and Remmers, K. (2007) *Photovoltaics for Professionals: Solar Electric Systems Marketing, Design and Installation*, Earthscan, London

Deutsche Gesellshaft Fur Sonnenenergie (DGS) (2007) *Planning and Installing Photovoltaic Systems: A Guide for Installers, Architects and Engineers*, Earthscan, London

Laughton, C. (2010) *Solar Domestic Water Heating: The Earthscan Expert Handbook for Planning, Design and Installation*, Earthscan, London

Deutsche Gesellshaft Fur Sonnenenergie (DGS) (2010) *Planning and Installing Solar Thermal Systems: A Guide for Installers, Architects and Engineers*, Earthscan, London
Henderson, S. and Roscoe, D. (2010) *Solar Home Design Manual for Cool Climates*, Earthscan, London
Stapleton, G. and Neill, S. (2011) *Grid-connected Solar Electric Systems*, Earthscan, London

Magazines

Renewable Energy Magazine: www.renewableenergymagazine.com
Renewable Energy World: www.renewableenergyworld.com
Home Power Magazine: http://homepower.com

Simulation tools

The following section on simulation tools is by no means comprehensive. For a comprehensive list of all simulation programs available, visit the US Department of Energy (DOE) Building Energy Software Tools Directory at http://apps1.eere.energy.gov/buildings/tools_directory/. Bear in mind that software is being updated all the time and it is advisable to be clear about what you require before purchasing.

bSol

bSol simulates the thermal behaviour of a building over one year, considering its specific climate and the surrounding topography and architectural parameters, such as orientation of the building, thermal insulation, size of openings, choice of glazing, ventilation (with or without heat recovery) or thermal inertia. Within a building, bSol supports two different climate zones (e.g. heated and unheated zones). It can take account of the components of the building's thermal mass and relies on climate data from METEONORM (air temperature, diffuse global radiation, elevation of the sun, azimuth, etc.).

bSol is developed by the University of Applied Sciences Western Switzerland (HEVs), competence group on energy in Sitten and information is available at www.bsol.ch.

DAYSIM

DAYSIM is for architects, engineers and researchers and aims to assess the daylight contribution in a building during the design stage. It can calculate illuminance distributions under all sky conditions in a year (dynamic simulation) and predict the incident radiation on a PV panel or a thermal collector but the program is not suited for sizing these components. The amount of solar radiation at the building site over the course of the year must be known and entered in the form of TRY files. DAYSIM directly imports .epw files and extracts the information required for an annual daylight simulation (global horizontal radiation and diffuse radiation or direct and diffuse radiation). DAYSIM does not specifically calculate solar gains through windows or allow sizing of PV and thermal installations, although some of the results might be used with some creativity in the process of planning other solar aspects.

DAYSIM is a Radiance-based daylighting analysis tool that has been developed at the National Research Council of Canada in collaboration with the Fraunhofer Institute for Solar Energy Systems (ISE) in Germany. Windows and Linux versions of DAYSIM can be downloaded free from www.nrc-cnrc.gc.ca/eng/projects/irc/daysim/download.html.

EnergyPlus

EnergyPlus models heating, cooling, lighting, ventilating and other energy flows as well as water in buildings. It includes many innovative simulation capabilities, such as time steps of less than an hour, modular systems and plant integrated with heat balance-based zone simulation, multizone air flow, thermal comfort, water use, natural ventilation and PV systems. EnergyPlus is a stand-alone simulation program without a 'user friendly' graphical interface. As a result, various software can be bought that provides such a convenient interface, including DesignBuilder (see below). EnergyPlus comes with real-time weather data in the correct format that can be input for any location.

EnergyPlus is provided by the US government and is available at http://apps1.eere.energy.gov/buildings/energyplus.

DASTPVPS

The DASTPVPS (Design And Simulation Tool for PhotoVoltaic Pumping Systems) simulation and design program can serve technical planners as an instrument for system design and for checking the performance of PV-powered pumps. It is in the form of an Excel spreadsheet. Operational data analyses indicate that a properly designed standardized PVP system can correspond well to the predictions of this program.

DASTPVPS has been developed by the German Universität der Bundeswehr in Munich and is available at www.ibom.de/dastpvps.htm

DesignBuilder

Design Builder (DBS) is a comprehensive user interface to the EnergyPlus (E+) dynamic thermal simulation engine providing an easy-to-use yet powerful 3D modeller. It simplifies the data input for E+ and allows setting predefined parameters for easily checking building energy demand, CO_2 emissions as well as lighting and comfort performance. It also includes the calculation of heating and cooling system sizes. For each of the building zones, the calculated loads can be obtained for heating and cooling the zone depending on lowest or highest outside dry-bulb temperature from the input weather data. It is possible to simulate, for example, wind conditions around the building as well as thermal behaviour in larger rooms of the building design.

Passive solar gains through windows and glass facade elements are calculated. DSB also includes daylighting models to simulate savings in electric lighting by using lighting control systems.

DSB at www.designbuilder.co.uk, has been developed by DesignBuilder Software Ltd.

Design Performance Viewer

Design Performance Viewer (DPV) allows architects and planners to determine the energy and exergy consumption of a building at the design stage. (Exergy is the energy cost of bringing a building to the same temperature as its surroundings). It is possible to test energy and exergy strategies by using different combinations of technical and construction systems, to create more sustainable buildings in terms of energy, economy and architecture. The implementation of the METEONORM climate data in upcoming releases as well as a link with the CRB cost database are planned.

DPV is currently being developed by Keoto AG, c/o ETH Zurich, Institute for Technology in Architecture at www.keoto.net.

Ecotect

Ecotect is a graphical building and environmental analysis tool. It can perform a large number of different analyses within a 3D model, including shadows and reflection, shading design, solar, lighting design, views and light, acoustic, thermal, and ventilation and air flow. The package can also create data concerning resource management and the environmental impact of building materials.

Ecotect was developed by Dr Andrew Marsh and Square One Research Limited. It is possible to register on the Ecotect website at www.ecotect.com and receive a free 30-day trial version.

Energy Design Guide II

Energy Design Guide (EDG) II supports the design of a building with low heating and cooling demand at a very early design stage. As the building is meant to set the boundaries for sizing the HVAC system, only the heating and cooling demand of the building (and not the way it is covered) is calculated. This strategy requires first minimizing the energy demand of a building and second, covering demands using efficient heating and cooling systems. It considers three factors:

- heat loss K [W/m^2K];
- solar temperature-correction factor γ [m^2K/W];
- time constant τ [s,h], defined through the dynamic thermal storage capacity.

EDG II has been developed by the Institute for Building Technology at ETH Zurich, by Professor Dr Bruno Keller and Stephan Rutz. Further details are available at www.energy-design-guide.ch.

ENERGIEplaner

ENERGIEplaner is relevant mostly to Germany and will help to investigate and establish the costs of building projects as a basis for cost calculations. The costs of a renovation may be compared to the reduction of energy costs in the building in different variations. The following energy consumptions are calculated:

- heating;
- domestic hot water;
- lighting;
- cooling;
- ventilation (air conditioning).

BKI ENERGIEplaner Polysun is an additional module, which offers a PV simulation with which it is possible to match the electrical gains of PV against primary energy consumed by the building.

ENERGIEplaner is provided by BKI, a subsidiary of the architectural associations of Germany.

ENERGY-10

ENERGY-10 is a design tool that analyses the energy and cost savings that can be achieved through more than a dozen sustainable design strategies. Hourly energy simulations help to quantify, assess and clearly depict the benefits of daylighting, passive solar heating, natural ventilation, well-insulated envelopes, better windows, lighting systems, mechanical equipment and more. It comes with a PV module that provides the ability to model and simulate the performance of a PV system that is either stand-alone or integrated with the building. A Solar Domestic Hot Water module provides hot water modelling capability. A library ('ASHRAELIB') is included defining constructions (wall, roof, window, etc.). It can help model overhangs and window sizes to calculate the optimum size and orientation of windows for any location.

ENERGY-10 is part of the Whole Building Design Guide, and is a program of the National Institute of Building Sciences; it is available at www.wbdg.org/tools/e10.php.

eQUEST

eQUEST is a building energy simulation tool. This software combines a building creation wizard, an energy-efficiency measure wizard, and a graphical reporting tool with a simulation 'engine' derived from the latest version of DOE-2. (DOE-2 is a command line program, for which the user creates input files, with the building description in DOE-2's building description language). It helps the designer address factors that would impact energy use, such as:

- architectural design;
- HVAC equipment;
- building type and size;
- floor plan layout;
- construction materials;
- area usage and occupancy;
- lighting system.

It calculates heating or cooling loads for each hour of the year, based on information about walls, windows, glass, people, plug loads and ventilation. eQUEST does provide a detailed calculation of passive solar heat gains but includes a PV module that supports active solar design. This makes it possible to determine voltage, angle, dimension and temperature between the cell and the back surface. However, it is not possible to model solar thermal systems in eQUEST. It can also predict the performance of skylights, lighting system controls and the effects of obstructions on daylight that can affect lighting system response.

eQUEST is a part of the DOE-2 software developed by James J. Hirsch & Associates in collaboration with Lawrence Berkeley National Laboratory. It is freeware and is available on the website of Energy Design Resources, a Californian foundation that offers a palette of decision-making tools and resources to design, build and operate more energy-efficient buildings: www.energydesignresources.com.

Green Building Studio

Green Building Studio (GBS) is an analysis tool that performs whole building energy analysis. The user specifies the building type and geographical location and there are built-in libraries for various simulation parameters (building construction, shading, internal heat gain, infiltration, schedule, HVAC system and equipment and utility rates). Users can evaluate how the building components affect energy use. This software also performs Leadership in Energy and Environmental Design (LEED) reporting for some credits about daylighting, water efficiency, energy and atmosphere.

LEED is an internationally recognized green building certification system, providing verification that a building or community was designed and built to increase energy savings, water efficiency, CO_2 emissions reduction, indoor environmental quality and stewardship of resources and sensitivity to their impacts. It is developed by the US Green Building Council (USGBC).

The Design Alternative feature helps users to determine the variations to a run that will improve energy efficiency. But GBS does not provide modelling tools and is for the US territory only.

GBS is a web-based energy analysis service supplied by Autodesk. Subscriptions provide an unlimited access to the web service including unlimited projects and unlimited runs.

IDA ICE

IDA ICE is building simulation software used for predicting and optimizing heating and cooling loads, energy consumption and thermal comfort in buildings. It calculates the dynamic interaction between the ambient climate, the building, the building system and the building occupants. Variables are resource consumption (energy, CO_2) and comfort values (temperature, operative temperature, hours with high temperature, humidity, CO_2-concentration, Fanger comfort values, etc.). Solar loads and shading calculations are performed, as well as active solar calculations. Component models for PV and solar thermal do exist but are not available at the user levels for the early planning phase.

IDA ICE is developed by the EQUA Simulation Technology Group AB in Stockholm, see www.equa.se.

Virtual Environment

Virtual Environment (VE) PRO is a bundle of modules that work with the same geometrical model within VE PRO. The modules can perform almost any aspect of building analysis from energy simulation using the heat balance method and sophisticated lighting analysis using Radiance to a simplified CFD module. The quantity of results that can be obtained from VE is vast; including comfort values, different radiant temperatures and so on.

VE is developed by Integrated Environmental Solutions (IES) in Scotland, and can be found at www.iesve.com.

PHPP

PHPP is a comprehensive package for helping designers produce buildings that meet the Passivhaus standard of only a 15kWh/m^2/annum energy requirement. It has been developed by comparing its modelling and predictions against hundreds of real passive house constructions. It offers:

- energy calculations (incl. R- or U-values);
- design of window specifications;
- design of the indoor air-quality ventilation system;
- sizing of the heating load;
- sizing of the cooling load;
- forecasting for summer comfort;
- sizing of the heating and domestic hot water (DHW) systems;
- calculations of auxiliary electricity, their primary energy requirements (circulation pumps, etc.), as well as projection of CO_2 emissions;
- verifying calculation proofs of KfW and Energy Conservation Regulations, Germany's energy efficiency building code (EnEV) (Europe);
- plus many more tools, e.g. a calculation tool to determine internal heat loads, data tables for primary energy factors, etc.;
- a comprehensive handbook.

Climate data sets for Eastern Europe, Austria, Switzerland and many European cities are included; the American version contains data for 40 US cities and imperial units. Users input measurements from their plans, material specifications, measurements and other data and can run the model to see the consequences. PHPP requires information on all areas, depths, angles and dimensions, including those of the frame, reveal and glazing and any shading outside, whether it is from buildings across the road, or trees, etc.

PHPP has been developed by Dr Wolfgang Feist at the Passivhaus Institut, Germany and can be purchased cheaply from www.passivhaus.org.uk, www.passivehouse.us and www.passivhaustagung.de.

Polysun

Polysun is used for the prediction of system profit ratio (early planning phase) and system optimization (detail planning). Polysun has a full featured solar simulation engine, including in-depth representation of thermodynamic and PV physics. In particular, it offers the combination of solar thermal, PV and heat pump applications. The program enables easy PV system sizing. Calculations are based on a dynamic simulation model and statistical weather data. As a result, it is a valuable tool for optimization and design, quality control and marketing support. With its solar thermal features, Polysun is widely used in retrofit applications. It offers automatic inverter matching as well as accurate yield prediction. It comes with a component catalogue database of PV modules, inverters, solar thermal collectors, heating units, etc. Shading analyses can be carried out, although in more complex shading situations PVsyst (see below) is more accurate.

Polysun is developed and sold by Vela Solaris AG, a privately-held Swiss corporation, a spin-off from the Institute for Solar Technology SPF at the University of Applied Science Rapperswil in Switzerland. Details from www.velasolaris.com.

PVsyst

PVsyst is a software package for the study, sizing, simulation and data analysis of complete PV systems. At preliminary design, PVsyst is used to pre-size systems. It also evaluates the monthly production and performances and carries out a preliminary economic evaluation of the PV system. At project design stage, the software generates detailed simulations computing hourly values and supports the user in the definition of the PV-field and the selection of the right components. It is possible to import a meteorological database as hourly values from METEONORM, Satel-Light, US TMY2 and HelioClim-2. It also has access to meteorological data as monthly values such as METEONORM, WRDC, NASA-SSE, PVGIS-ESRA and RETScreen. Output of solar geometry is possible (sun paths, incidence angles, etc.), various graphs of the components' behaviour (PV modules, batteries, pumps), hourly meteorological plots and calculations of irradiation on tilted planes. For stand-alone systems, the preliminary level calculates a size based on the required PV power and battery capacity. For pumping systems, the user may specify the water needs and pumping conditions. The preliminary level also estimates the required power of the pump and the PV array.

PVsyst is supplied by the ISE and the University of Geneva. A trial version is available for 15 days at www.pvsyst.com/5.2/index.php.

PV*SOL

PV*SOL is a program for the design and simulation of PV systems. It is possible to create a system using a wide range of modules and to determine its size with the roof layout facility.

After testing all relevant physical parameters, the program reveals the appropriate inverter and PV array configuration. It is possible to determine the capital value of the system, the electricity production costs and the amortization period. The calculations are based on hourly balances. The output of a PV system may be based on the characteristic curve for each of the PV modules contained in the database. Climate data for irradiation and air temperature is delivered for approximately 650 European locations. TRY and METEONORM format climate files for any location can be imported. Shading analyses can be carried out. If the monthly values for a particular location are not available, they can be obtained from the NASA website.

PV*SOL is supplied by Valentin EnergieSoftware GmbH, Berlin, Germany www.valentin.de; a free download is available at www.solardesign.co.uk.

Radiance

Radiance is a suite of programs for the analysis and visualization of lighting in design. It is able to predict internal illuminance and luminance distributions (including glare) in complex buildings or boundary spaces under arbitrary sky conditions or electrical lighting. Radiance uses ray tracing in a recursive evaluation of the luminance integral to a room. It also has an accurate engine for the calculation of solar irradiance on different surfaces (defined via a sensors matrix). The simulations in Radiance can be performed hour by hour. If users need annual simulations or long-term estimations, it is possible to use the DAYSIM program. Radiance does not include algorithms for the calculation of passive solar heat gains or sizing active PV or ST systems.

Radiance has been developed by Greg Ward at the Lawrence Berkeley National Laboratory and can be downloaded for free from radsite.lbl.gov/radiance/.

RETScreen

RETScreen evaluates the energy production and savings, costs, emission reductions, financial viability and risk for various types of renewable-energy and energy-efficient technologies. The software also includes product, project, hydrology and climate databases, a detailed user manual, and a case study-based college/university-level training course, including an engineering e-textbook.

There is no information about consideration of passive solar gains in RETScreen. In the rendering production options, PV production is included. However, it is the user who has to define how PV modules contribute to electricity production. It is the same with ST systems for hot water. So RETScreen considers active solar systems, but does not provide solar calculations.

RETScreen is a Clean Energy Project Analysis software developed by Natural Resource Canada through collaboration between many experts from industry, government and academia. It may be downloaded for free at www.retscreen.net/ang/home.php. It is a Microsoft Excel based application.

The RETScreen Software Solar Water Heating Model can be used anywhere in the world to evaluate the energy production and savings, costs, emission reductions, financial viability and risk for solar water heating projects. Projects can be any size, from solar collectors for small-scale domestic hot water applications to indoor and outdoor swimming pools for residential, commercial and institutional buildings and large-scale industrial applications and even aquaculture.

The software contains a database of typical daily hot water use for loads, such as houses, apartments, hotels and motels, hospitals, offices, restaurants, schools, laundry rooms and car washes. The software (available in multiple languages) also includes product, project and climate databases, and a detailed user manual. It is available free at www.retscreen.net/ang/g_solarw.php.

T*Sol

T*Sol is a simulation program for solar thermal heating systems, including hot water preparation and space and swimming pool heating. It is possible to simulate different configurations consisting of components of a database. Starting from standard components (i.e. 1200 different collectors), parameters can be changed according to the individual system. An automatic design assistant helps to optimize the design/dimensioning of collector area and storage tank volume. The influence of partial shading by the horizon and other objects such as trees or buildings can be included as well as different consumption patterns. Besides the standard evaluation values for solar thermal systems (produced energy [hot water and space heating] of the solar system, efficiency, solar fraction, etc.), the program also calculates economic efficiency. You can use the same climate data as for PV*Sol, and actual solar calculation is similar to PV*Sol.

A supplier of T*Sol is Valentin EnergieSoftware GmbH, Berlin, Germany [www.valentin.de]; a free download is available at www.solardesign.co.uk.

THERM

THERM can help to determine total window product U-values or R-values and Solar Heat Gain Coefficients for two-dimensional modelling of heat transfer effects in windows, walls, foundations, roofs, doors, appliances and other products where thermal bridges are of concern. THERM permits evaluation of a product's energy efficiency and local temperature patterns, condensation, moisture damage and structural integrity. It can produce graphic visualizations such as the one in Figure 10.10 of cross-sections, isotherms, colour infrared heat transfer cross-sections and more. THERM's heat-transfer analysis allows the evaluation of a design's energy efficiency and local temperature patterns.

THERM has been developed at Lawrence Berkeley National Laboratory. It is available at http://windows.lbl.gov/software/therm/therm.html

VisualDOE

VisualDOE is an energy simulation program used to accurately estimate the performance of building design alternatives. VisualDOE covers all major building systems including lighting, daylighting, HVAC, water heating and the building envelope. Among the range of simulation results are electricity and gas consumption, electric demand and utility cost. Hourly results are available for

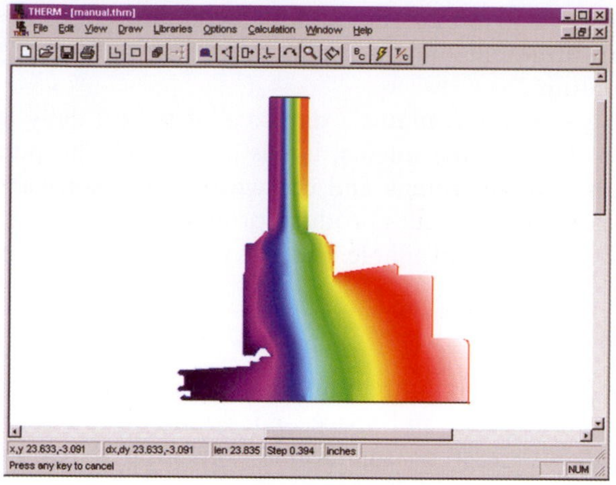

Figure 10.10 A screen grab taken from THERM software. It shows a cross-section through a window frame. The bluer colours represent cold and the redder colours hot. It indicates the flow of heat through the frame showing how effective it is at keeping heat inside.

a detailed analysis. The software also offers energy calculations for LEED certification. It is possible to conduct simulations and report on absorption of solar radiation by the roof and exterior walls transmission, absorption of solar radiation by windows and internal gains.

Since PV energy can be added or subtracted pretty much independently of the rest of the building energy use and loads, the total kWh contribution can be analysed using PVWATT. It only requires some very basic inputs such as square feet, AC-DC derating, location (latitude) and mounting declination angle. PVWATT is especially suitable for obtaining the full annual energy contribution, and not the hourly increments.

VisualDOE is supplied by Architectural Energy Corporation. It is possible to download a trial version from www.archenergy.com/products/visualdoe/.

Detailing for building design and construction

Detailing to eliminate heat losses that occur at the junctions between building elements and around openings is available online for free at the UK Energy Saving Trust's Enhanced Construction Details (ECDs) minisite: www.energysavingtrust.org.uk/business/Business/Housing-professionals/Interactive-tools/Enhanced-Construction-Details.

Solar thermal

IEA's Solar Heating and Cooling Programme: www.iea-shc.org.

Solar cooking

NASA's Surface Meteorology and Solar Energy Database can reveal if there is enough sunshine to use solar cooking in any part of the world. Levels of 4kWh/m^2 and above are suitable for solar cooking. This database is available at: www.eso.org/gen-fac/pubs/astclim/espas/world/ION/ion-pwv.html.

Solar Cookers International (SCI): http://solarcooking.org

Index